Jules A Colas

Poole Bros' Celestial Handbook

Companion to Their Celestial Planisphere

Jules A Colas

Poole Bros' Celestial Handbook
Companion to Their Celestial Planisphere

ISBN/EAN: 9783744691970

Printed in Europe, USA, Canada, Australia, Japan

Cover: Foto ©Andreas Hilbeck / pixelio.de

More available books at **www.hansebooks.com**

POOLE BROS'

CELESTIAL HANDBOOK

COMPANION TO THEIR

CELESTIAL PLANISPHERE.

COMPILED AND EDITED BY

JULES A. COLAS.

ILLUSTRATED.

CHICAGO:
1892.

CONTENTS.

INTRODUCTORY.

We can not do better in introducing our Handbook than to quote from Richard A. Proctor's "Half Hours with the Stars" the following lines:

"It is very easy to gain a knowledge of the stars, if the learner sets to work in the proper manner, but he commonly meets with a difficulty at the outset of this task. He provides himself with a set of the ordinary star-maps and then finds himself at a loss how to make use of them. Such maps tell him nothing of the position of the constellations *on the sky.* If he happen to recognize a constellation, then, indeed, his maps, if properly constructed, will tell him the names of the stars forming the constellation, and, also, he may be able to recognize a few of the neighboring constellations. But when he has done this he may meet with a new difficulty, even as respects this very constellation. For if he look for it again some months later, he will neither find it in its former place nor will it present the same aspect, if, indeed, it happen to be above the horizon at all."

The object of our planisphere is to show the aspect of the heavens as it really is *at any moment* by setting it properly in position for the time of observation, and no matter how large a constellation is it can be found very easily by simply turning the movable part a little one way or the other; its great advantage over the ordinary star-maps is the facility afforded to the student to find the position and the character of the constellation that he is looking for, at any time, and to save the trouble of trying to find one which may not be above the horizon at the moment.

It contains nearly all the stars visible to the naked eye from our latitude; also the principal curiosities accessible to common telescopes.

The immense distance which separates the stars from our solar system leaves no room to doubt that they are *suns like our own.* Seen from Sirius, the earth, the moon and the sun will appear as a spot only, and the thickness of a hair will eclipse them entirely. The earth's motion around the sun brings us 184,000,000 miles nearer to certain stars than they are six months after, but this distance, which seems so great, does not increase the apparent magnitude of any of them.

The planets such as Venus, Jupiter, Saturn, etc., when seen through the telescope, present a certain diameter, and on their surface some details can be seen—they appear so much larger and nearer to us according to the power of the instruments. This is not the case with the stars; they are so far from us that the most powerful instruments fail to produce an apparent diameter; they only appear more brilliant, but not larger; on the contrary, the stronger the instrument the smaller the apparent diameter; consequently it must be well understood that what is called the magnitude of a star is *not its size,* but its brightness; a star invisible to the naked eye may be *very near* to us; *1830 Groombridge,* for example (see Ursa Major notes), and a bright star like *Rigel,* in the Constellation Orion, which offers no parallax, must be very far from us.

The most conspicuous are called stars of the first magnitude; the next in brilliancy are stars of the second magnitude; the next, of the third magnitude, etc.

.The stars comprised between the first and the sixth magnitude are visible to the naked eye on a clear night; but a telescope of large power reveals the stars down to the twentieth magnitude.

From one magnitude to another there is quite a difference in brightness, and the astronomers have divided them into tenths; consequently a star of 2.9 magnitude is very near like a star of the third magnitude; but from one class to another there is the same difference in brightness.

The following table is taken from Mr. G. Hermite's article on the "Determination of the Number of Stars of Our Universe:"

Magnitudes.	Total number of stars of each magnitude.	Number of stars of each magnitude necessary to equal one star of first magnitude.	Number of stars of first magnitude necessary to produce the same amount of light as total number of stars of corresponding magnitude.
1	20	1	20
2	59	2.56	23.046
3	182	6.55	27.771
4	530	16.78	31.59
5	1,600	42.95	37.243
6	4,800	109.96	43.65
7	13,000	281.68	46.15
8	40,000	721.09	55.47
9	100,000	1,846.90	54.17
10	400,000	4,726.75	84.64
11	1,000,000	12,008.9	82.65
12	3,000,000	30,870.6	97.18
13	10,000,000	79,028.8	126.53
14	30,000,000	202,314	148.28
15	*90,876,411	517,923	*175.4
16	272,629,233	1,323,513	206.9
17	817,887,699	3,388,175	241.3
18	2,453,663,097	8,673,738	282.8
19	7,360,989,291	22,204,745	331.5
20	22,082,967,873	56,844,147	388.8

* The figures from the first to the fourteenth magnitude are given from direct observations; from the fifteenth to the twentieth magnitude the figures have been calculated.

The first column represents the magnitudes; the second column, the number of stars of each magnitude; the third column, the number of stars of each magnitude necessary to produce the same luminous intensity that one star of the first magnitude does; the fourth column, the number of stars of the first magnitude necessary to produce the same amount of light as the total number of stars of the corresponding magnitudes; for instance, the 1,600 stars of the fifth magnitude produce the same light as about 37 of the first magnitude, and 22,082,967,873 of the twentieth magnitude produce about the same light as 389 of the first magnitude. It requires about 43 stars of the fifth magnitude to produce the same light as one star of the first magnitude, and of the twentieth magnitude it requires 56,845,000 stars to produce the same result.

According to W. Herschel the full moon gives the same light as 27,408 stars of the first magnitude, and the total amount of light produced by all the stars is equal to only one-tenth of the light produced by the full moon, or equal to the light of 2740.8 stars of the first magnitude. By progressive calculations Mr. G. Hermite came to the conclusion that the total number of stars of *our universe* is equal to 66,000,000,000 (66 billions). (Revue d'Ast., 1886; pages 406-412.)

In comparing the photometrical measures of John Herschel, Langier, Secchi, Seidel and Trépied, the twenty stars of the first magnitude came in the following order:

STARS.	INTENSITY OF LIGHT.	MAGNI- TUDES.	STARS.	INTENSITY OF LIGHT.	MAGNI- TUDES.
Sirius	400	0.25	Aldebaran	46	1.6
Canopus	200	0.5	Antares	45	1.6
a (Alpha) Centauri	100	1.0	β (Beta) Centauri	45	1.6
Arcturus	75	1.2	a (Alpha) Crux	44	1.7
Vega	72	1.2	Altair	43	1.7
Rigel	68	1.3	Spica	41	1.8
Capella	63	1.3	Fomalhaut	41	1.8
Procyon	58	1.4	β (Beta) Crux	40	1.8
Betelgeuse	50	1.5	Regulus	40	1.9
Achernar	48	1.6	Pollux	38	1.9

This plainly shows that the light coming from Sirius is four times stronger than the light of *a (alpha)* Centauri, and ten times stronger than the light of Regulus.

Father Secchi divided the stars at first into three classes ; later, however, into four, according to the nature of their spectra.

FIRST CLASS — WHITE STARS.

The white stars offer a continuous spectrum (Fig. I), in which appear several fine black lines. The great extension of the blue and violet portions of the spectrum indicates a *high temperature ;* the lines produced by the vapors of iron, magnesium, sodium, etc., are very faint and hard to distinguish. On the contrary, the *four lines of hydrogen* are very prominent: they are the lines C, in the red; F, in the greenish-blue; G', in the blue, and h, in the violet. The white stars form about 60 per cent of the total number of stars, among them Sirius, Vega, Rigel, Altair, Regulus, 75 Pegasi, etc.

SECOND CLASS — YELLOW STARS.

The spectra of the yellow stars (Fig. II) are characterized by a great number of black lines well defined, and corresponding to the vapors of our different metals. The lines of iron, magnesium, cobalt, chrome, sodium, etc., are identified without doubt; the blue and violet portions are not so intense as in the white stars, and this accounts for their yellow coloration. The lines of hydrogen exist also, but less defined. The yellow stars form about 35 per cent of the total number of the stars. Our Sun, Aldebaran, Capella, Arcturus, *a (alpha)* Bootis, 70 Ophiuchi, etc., belong to this class. These stars have most likely commenced to cool off.

THIRD CLASS — REDDISH STARS.

In the spectra of this class, the blue and violet portions are very feeble; the lines of hydrogen are generally absent, and it presents, near the ordinary lines, some lines of absorption that give to the spectrum a fluted appearance (Fig. III). These bands or flutings always shade from the blue towards the red. These reddish stars form about 5 per cent of the total number of stars; among them are *a (alpha)* Herculis, Antares, Betelgeuse, R. Leonis, U. Virginis, etc. These stars are most likely still further advanced in the cooling process. There are also some telescopic stars of a very dark red ; they are most likely suns in an advanced state of oxidation.

FOURTH CLASS.

This includes a small number of stars having also a fluted spectrum, but with the sharp edges of the bands turned towards the red. Most stars of this class show bright lines, *e. g. 152 Schjellerup.*

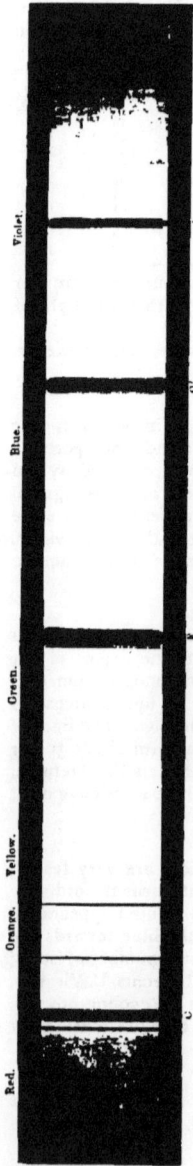

Fig. I.—Spectrum of Sirius. (White Stars.)

Fig. II.—Spectrum of the Sun. (Yellow Stars.)

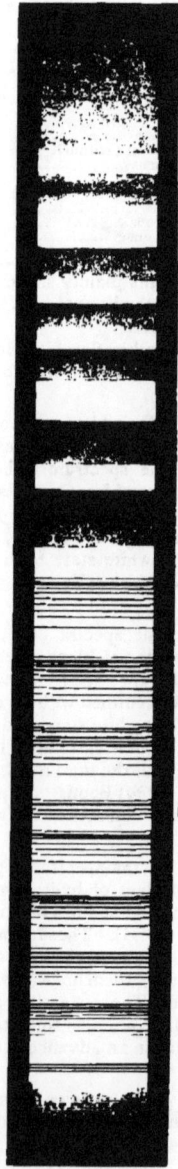

Fig. III.—Spectrum of α (alpha) Herculis. (Reddish Stars.)

When you look through a telescope and see a star near another, it is not a proof that it is a *double star;* it may be an optical illusion, and the two stars are sometimes at a great distance from each other and do not form a *physical system.* It is only by several years of patient and careful observations that they are recognized as having the same motion or as revolving around each other ; therefore it is only those which sometimes show two, three, four or more stars in close proximity to each other that we call *double stars, triple stars, quadruple stars,* etc.

Of this class of stars those revolving around each other in regular orbits are known as *binary stars, trinary stars,* etc.

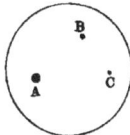

Let Figure IV represent a triple star as seen with a telescope. A, B and C are called the *components,* B and C are the *companions* of A.

A great number of stars are subject to periodical changes of brilliancy and are called *variable stars.* The *temporary* or *irregular stars* are those that have occasionally burst forth in the heavens with a brilliancy far surpassing the stars of first magnitude, in some instances remaining thus for a short period and then gradually fading away.

Fig. IV.

The *fixed stars* are not all of them stationary; in fact, very few, if any, are fixed. In comparing their location at different times it was found that a great number of them have a *proper motion* in the heavens; we have noted only those which offer the greatest motion.

In many places in the heavens there appear some faint luminous spots shining with a generally white, pale light; they are called *nebulæ ;* some can be distinguished with the naked eye, but most of them require the aid of the telescope.

The *clusters* are *nebulæ* which are *easily resolved into stars.*

The *nebulæ proper* offer no appearance of stars, and are supposed to be masses of nebulous matter; in some of them gases have been found through the spectroscope and their nature determined.

The *planetary nebulæ* are those offering an appearance similar to the planets; they are round, of equal light, and their outline is perfectly defined; some of them have been resolved into stars, but it requires telescopes of great power.

The *stellar nebulæ* are much condensed at the center, and offer the appearance of a star seen through the surrounding nebulous mass.

The *nebulous stars* are stars distinctly seen to be such, surrounded by their nebulous atmospheres; they are very likely stellar nebulæ in a more advanced state of progressive condensation.

If two straight lines AB, AC (Fig. V) meet each other at A, they form what is called an *angle,* and the distances between AB and AC at *any two points* are proportionate to the distance of these two points from the intersection A.

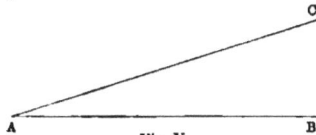

Fig. V.

An angle does not change with the distance; one degree on a sheet of paper or one degree on the sky is always one degree; in a circumference of one inch in diameter, or one mile in diameter, or one million of miles or ten millions of miles in diameter, a certain angle drawn from the center always intersects the same proportion of the circumference.

One degree is the $\frac{1}{360}$ part of a circumference, and is marked 1°
One minute is the $\frac{1}{60}$ part of a degree, and is marked 1′
One second is the $\frac{1}{60}$ part of a minute, and is marked 1″
3° 45′ 31″ means 3 degrees, 45 minutes and 31 seconds.

An angle of 1 degree corresponds to a distance of...................... 57
  "   ½ degree, or 30 minutes, corresponds to a distance of... 114
  "   1½ degree, or 6 minutes, corresponds to a distance of..... 570
  "   1 minute corresponds to a distance of.................... 3,438
  "   ½ minute, or 30 seconds, corresponds to a distance of... 6,875
  "   20 seconds corresponds to a distance of 10,313
  "   10 seconds corresponds to a distance of 20,626
  "   1 second corresponds to a distance of 206,265
  "   0″9 corresponds to a distance of......................... 229,183
  "   0″8   "   "   "   257,830
  "   0″7   "   "   "   294,664
  "   0″6   "   "   "   343,750
  "   0″5   "   "   "   412,530
  "   0″4   "   "   "   515,660
  "   0″3   "   "   "   687,500
  "   0″2   "   "   "   1,031,320
  "   0″1   "   "   "   2,062,650
  "   0″0   "   "   "   immeasurable.

The above table indicates the distances corresponding to certain angles, and when we know the angle, we know the distance corresponding to it.

We can not insist too strongly on the importance of a clear understanding of what an angle is, for the very reason that it is *the base of all astronomical observations.*

One degree is one inch seen at the distance of   57 inches.
One minute is one inch seen at the distance of   3,438 inches.
One second is one inch seen at the distance of 206,265 inches.

If we were using one foot or one mile or any other measure instead of an inch the same rule would apply.

In the case of the stars the standard measure is one-half the diameter of the earth's orbit, or 92,000,000 miles.

The *parallax of a star* is the greatest angle that can be subtended by the radius or half the diameter of the earth's orbit as seen from the star in question. In Figure VI let S be the sun, ES the radius of the earth's orbit, and A the position of a star, the angle EAS is the parallax.

As the mean radius of the earth's orbit is equal to 92,000,000 miles, when we know the parallax of a star, it is very easy to find its distance; if a star has a parallax of 0″5, for example, in referring to the table above it shows that the distance of this star from us is 412,530 times 92,000,000 miles, or 37 trillions 952 billions 760 millions of miles.*

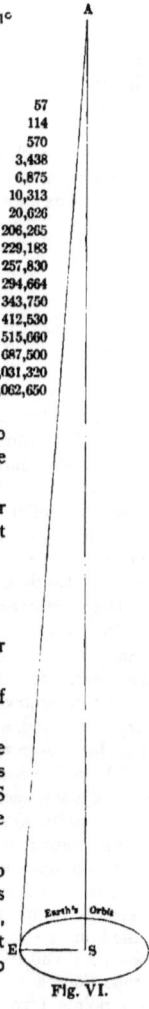

Fig. VI.

---

* NOTE.—The distance of a star is obtained by this formula:
$$\frac{206,265 \; R.}{P}$$
R is the radius of the earth's orbit and P the parallax of the star.
The number of years necessary for the light to come from a star to us is given by this formula:
$$\frac{3.262}{P}$$
3.262 is the number of years corresponding to a parallax of one second, and P is the parallax of the star.

Let Fig. VII represent a celestial sphere and S a star; the *declination* is the distance, SB, from the star to the equator measured on the great circle, PBP'; it is *north declination* or *south declination* according to the position of the star in regard to the equator. If the distance SB = 19° 45', declination +19° 45' indicates that the star is 19 degrees 45 minutes north of the equator; declination —19° 45' indicates that the star is 19 degrees 45 minutes south of the equator.

The *right ascension* of a star is the distance, OB (measured on the equator), between the vernal equinox, O, and the declination circle, PBP'; it is sometimes expressed in degrees, but generally in *time;* the right ascension is always measured toward the EAST according to the apparent motion of the sun among the constellations;if B is at 2 hours and 15 minutes from O, the position of the star S is exactly determined, and is marked thus: R. A.=2h. 15m.; Decl.= +19° 45'.

The sun does not cross the equator at the same point every year.

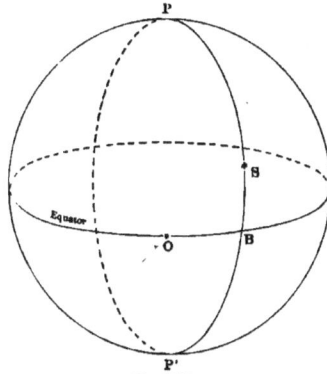

Fig. VII.

Hipparchus, 127 years B. C., in observing Spica and Regulus and in comparing the positions of these stars with the positions given by the Babylonian astronomers 2,120 years B. C., noticed a difference of 28 degrees for 2,000 years; the equinoctial point shifts 50" 3 along the equator every year, and all the stars move from west to east, describing a complete circle in a period of 25,765 years.

This motion is known as the *precession of the equinoxes.* Two thousand years ago the sun was in the constellation Aries at the time of the vernal equinox; now, on the 21st of March, it is in the constellation Pisces.

The right ascension of a star, or the distance of a star to the meridian of the vernal equinox, is augmented a little more than three seconds in time every year by the precession.

The spinning of a top gives a very fair representation of the precession of the equinoxes. The axis of the top, Fig. VIII, describes a cone, ABC, and the point B, which represents the celestial pole, describes a complete circle in 25,765 years; that is why the celestial pole changes every year; why α (*alpha*), in the constellation Ursa Minor, is now the pole star and will be nearest to the pole in the year 2105.

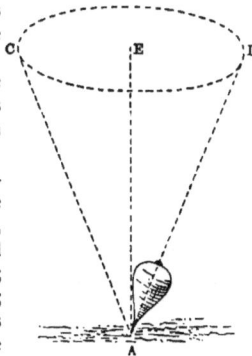

Fig. VIII.

In Fig. IX the plain line represents the classical circle described by the celestial pole by the effects of the precession; but as the solar system is going toward the constellation of Hercules, M. Flammarion has indicated a dotted spire which represents the *probable position* of the celestial pole, due to this motion, for a period of 28,000 years. (Revue d'Ast., 1886; page 401.)

In looking at this diagram it will be seen that ◦ (*alpha*), in constellation Draco, was polar star 2,700 years B. C.; ◦ (*iota*), of the same constellation, was polar star 4,500 years B. C.; ϒ (*gamma*), in Cepheus, will be polar star in about 2,600 years from now; ◦ (*alpha*), of the same constellation, will be the pole star in 7500 A. D.; it will be ◦ (*delta*), in constellation Cygnus, in 11300 A. D.; and the nice star Vega was polar star 12,200 years B. C. and will be in 13,000 years from now.

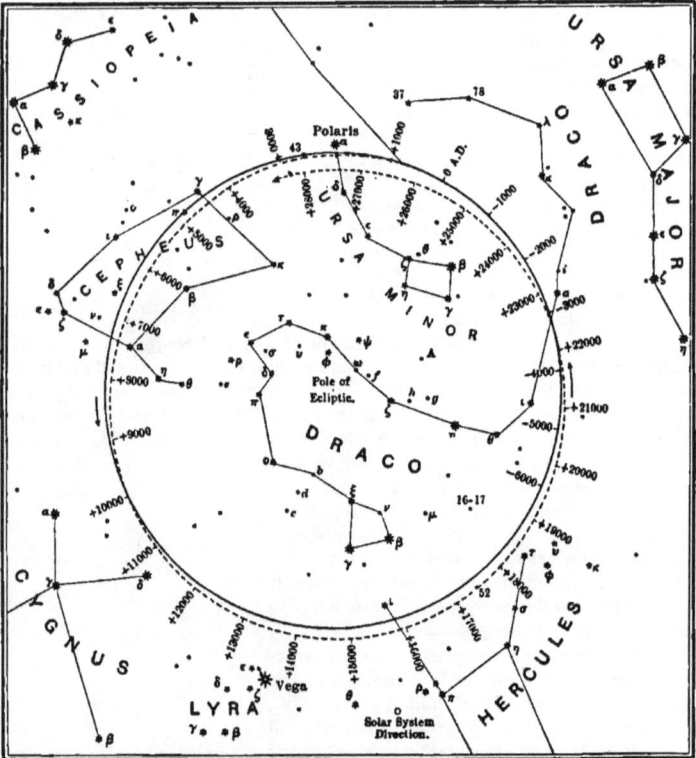

Fig. IX.—Diagram showing the position of the Celestial Pole for 6,000 years B. C. and 28,000 years A. D.

The *pole of ecliptic* is in the constellation Draco.

In our notes of each constellation will be found a short explanation of all curiosities, which will be of great interest to beginners; also how to find them when they do not appear on our planisphere.

We have noted the principal double stars with the distance and magnitude of their components; also the triple, quadruple and multiple stars in the same way.

The time of revolution of binaries is also given from the latest authorities; the parallaxes of the nearest stars and their distances to the earth have also received particular attention; the names of observers and the dates of observation have been given whenever accessible.

The temporary stars have also been described in the constellations in which they appeared, and their changes of magnitude and their position given as far as possible.

Our catalogue for each constellation is the same as the general catalogue of C. Flammarion, "Les Etoiles;" it contains *all the stars visible to the naked eye;* the stars of the sixth magnitude having a Greek or Latin letter; the principal variables; the stars whose distances have been found, and those remarkable for their colors; the principal double, triple, quadruple, etc., stars; also the principal clusters and nebulæ. For the double stars Mr. Burnham's catalogues were consulted.

Bayer's Greek letters have been given the preference; next, Flamsteed's letters, thus: Fl. 5; also Piazzi's Horal numbers, thus: P. XIV, 260; some stars of the British Association Catalogue also appear, thus: B. A. C. 1800; when other catalogues have been used the names of the authors appear. Hev. is for Hevelius; Radcl. for Radcliffe; x for Struve; ox for Otto Struve; Lal. for Lalande; Lac. for Lacaille.

Dbl. double; bin. binary; trip. triple; trin. trinary; qdl. quadruple, etc. V after the magnitude indicates the maximum of magnitude of a variable; M. before a nebula refers to Messier's catalogue; H. IV, 8, means Herschel's nebula number 8, volume IV; *neb.* nebula; *cl.* cluster.

The colors of the stars are denoted thus: Orange, org.; yellow, yel., etc.

When we say for example, *θ (theta)* Virginis is triple; magnitudes 4.5-9 and 10; distances, 7″ and 65″, it means that the star *θ (theta)* seen through the telescope consists of three stars; the first one of the 4.5 magnitude; the second of the 9th magnitude, and the third of the 10th magnitude; and that the star of the 9th magnitude is at a distance corresponding to seven seconds from the star of the 4.5 magnitude; and the star of the 10th magnitude is at a distance corresponding to one minute and five seconds from the star of the 4.5 magnitude.

The diagrams of the double or multiple stars are all made to the regular scale of one-fiftieth of an inch for one second, when a particular scale does not appear, and are drawn as they appear through a refracting telescope, which inverts the object.

The boundaries of the constellations are somewhat arbitrary; they are simply indicated to easily find the stars of the catalogue referring to each constellation.

R. A. Proctor said that if the constellations were entirely removed from the celestial atlases very little inconvenience would follow; here we quote his own words: "*Astronomy as an exact science would, in my opinion, gain greatly by the removal of the constellations, though I must admit that so far as popular astronomy is concerned I should be sorry to see the foolish old figures removed.*"

We entirely agree with the eminent astronomer, but in the interest of our readers we reproduce at the end of our Handbook a planisphere showing all the historical, mythological and old and new constellations, more as a curiosity than for utility.

CONSTELLATIONS NORTH OF THE ZODIAC.

URSA MINOR.

This constellation, now called "The Little Bear," was known as the "Phœnice" because it was the guide of the Phœnicians during their excursions and travels in the Mediterranean Sea; they used to call it *Cynosura* or *Dog's Tail*. It is commonly called "The Little Dipper." It is supposed to have been introduced by Thales, in the 7th century B. C.; it is mentioned by Eudoxus and Aratus. *a (Alpha)* is the Polar Star; *β (beta)* and *γ (gamma)* are called "The Guardians." The stars of this constellation never set, consequently they can be seen from our latitude every day of the year.

DESIGNA-TION.	MAGNI-TUDE.	POSITION			DESIGNA-TION.	MAGNI-TUDE.	POSITION		
		R. A. 1880 h. m.	DECL. ° '				R. A. 1880 h. m.	DECL. ° '	
a dbl.	2.0	1.15	+88.40		Fl. 5	4.8 red	14.29	+76.15	
β dbl.	2.2 red	14.51	74.39		Fl. 2	5.0 yel.	0.52	85.37	
γ	3.0	15.21	72.16		Fl. 4	5.4 org.	14.11	78. 6	
δ dbl.	4.3	18.11	86.37		Fl. 11	5.8	15.19	72.14	
ε dbl.	4.5	16.54	82.14		P. XIV, 260	5.2 red	14.56	66.25	
ζ dbl.	4.5	15.48	78.10		P. XIII, 109	5.6	13.24	73. 2	
η	5.0	16.21	76. 2		5058, B.A.C.	5.6	15.14	67.54	
θ	5.7 red	15.35	77.45		18 π dbl.	6.5	15.36	80.51	

NOTES.

a (Alpha) Polaris—Double; magnitudes 2.9 and 9.5; distance, 18".6; yellow and blue; somewhat difficult pair; it is now at 1° 20' from the pole; will be, by the effect of precession of the equinoxes, nearest to the pole in A. D. 2105 (see Fig. IX, page xii). The parallax of this interesting star has been obtained many times from Lindman, in 1841, to Pritchard, in 1888, and the average adopted gives it as 0".089 (Revue d'Ast., Dec., 1889).

AUTHORS.	PARALLAX.	AUTHORS.	PARALLAX.
Lindman, 1841............	0".144 ±0".056	Peters, 1853................	0".067 ±0".012
Peters, 1853........	0".075 ±0".036	Lindhagen, 1853.................	0".025 ±0".018
Peters, 1853.............	0".172 ±0".027	L. de Ball.....................	0".015
Peters, 1853.	0".147 ±0".030	Pritchard, 1888..................	0".075 ±0".015

This parallax represents 2,318,000 times the distance of the earth from the sun or 210 trillions of miles; the light traveling with a velocity of 187,500 miles per second would take no less than 36 years to reach us; a fast train going at the rate of 60 miles per hour would have to run without stopping for more than 479 millions of years.

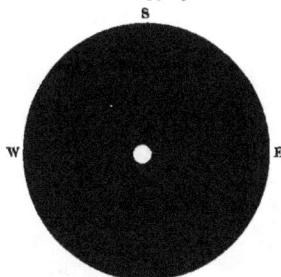

Fig. 1.—Double Star α

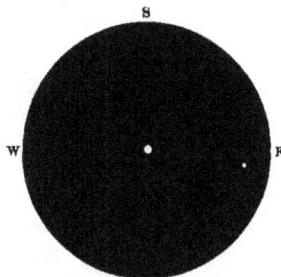

Fig. 2.—Double Star π

π (Pi)—Double; magnitudes 6.5 and 7.5; distance, 30"; yellow and blue; very easy pair.
γ (Gamma)—A person with good eyesight can see the star 11 at 57' (minutes) from γ (gamma).
δ—Double; magnitudes 4.8 and 11; distance, 45"; the little one is difficult to be seen.
Anonyma—Double; magnitudes 7.5 and 9; distance, 2"; it is the nearest to the pole.
R. (Cephei) -Variable; named as 23 Cephei by Hevelius; it varies from the 5th to the 10th magnitude in 365 days (see between a (alpha) and δ (delta).)

β *(Beta)Kochab*—Double; magnitudes 3 and 11; distance, 165″.   } The companions are small and
ζ *(Zeta)*—Double; magnitudes 4 and 11; distance, 310″.   } require instruments of good
ε *(Epsilon)*—Double; magnitudes 4 and 12; distance, 41″.   } power to separate them.

DRACO.

This constellation appeared already during the time of Eudoxus, of Cnidus, 4th century B. C., and represents the Dragon, which was the guardian of the golden apples of the garden of the Hesperides; according to others it was the monster killed by Cadmus. It has been introduced between the two Bears most likely to fill the space left between them; it is circumpolar, consequently visible all the year round.

DESIGNA-TION.	MAGNI-TUDE.	POSITION		DESIGNA-TION.	MAGNI-TUDE.	POSITION	
		R. A. 1880 h. m.	DECL. ° ′			R. A. 1880 h. m.	DECL. ° ′
α dbl.	3.3	14. 1	+64.57	χ dbl.	4.0	18.23	+72.41
β	2.9 yel.	17.28	52.23	ψ dbl.	4.7	17.44	72.13
γ	2.4 org.	17.54	51.30	ω	5.1	17.38	68.48
δ dbl.	3.0 yel.	19.13	67.27				
ε dbl.	4.4	19.48	69.58	15 A	5.3	16.28	69. 2
ζ	3.1	17.08	65.52	39 b, trip.	5.0	18.22	58.44
η dbl.	2.9	16.22	61.47	46 c, dbl.	5.3	18.40	55.25
θ dbl.	3.4	16. 0	58.53	45 d	5.0	18.30	56.57
ι dbl.	3.3 yel.	15.22	59.23	64 e	6.0	20. 1	64.23
κ	3.4	12.28	70.27	27 f	5.4	17.32	68.13
λ dbl.	3.6 red	11.24	70. 0	18 g	5.3	16.40	64.49
μ bin.	5.5	17. 3	54.38	19 h	5.3	16.55	65.16
ν dbl.	4.0	17.30	55.15	10 ξ, dbl.	5.0 org.	13.48	65.19
ξ	3.9 yel.	17.51	56.53	40 dbl.	5.4	18.12	79.59
ο dbl.	4.8	18.49	59.14	17 dbl.	5.8	16.33	53. 8
π	4.9	19.20	65.29	P. IX, 37	4.3	9.11	81.52
ρ	5.0	20. 2	67.31	P. X, 78	5.0 org.	10.22	76.20
σ	5.4	19.33	69.27	R	G.V.	16.32	67. 0
τ	5.0	19.18	73. 8	*	6.5 red	19.26	76.20
υ	5.2	18.56	71. 8	17415 (Eltzen	8.0	17.37	68.28
φ	4.3	18.23	71.16	H. IV, 37	neb.	17.58	66.38

NOTES.

The three first stars are named: α *(alpha)* Thuban, β *(beta)* Alwald and γ *(gamma)* Etanin.

ν *(Nu)*—Double; magnitudes 4.7 and 4.7; distance, 62″; very easy pair; an opera glass will separate them. Mr. Belopolsky, in 1888, found a parallax =0″32±0″076, and another 0″28±0″088.

ο *(Omicron)*—Double; magnitudes 4.7 and 8.5; distance, 32″; gold-yellow and lilac; nice pair; beautiful contrast.

γ *(Gamma)*—Mr. Auwers, in 1869, obtained for the parallax of this star, 0″.092±0″.070. It has a companion of 13th magnitude, discovered by Mr. Burnham, at Chicago; distance, 21″.

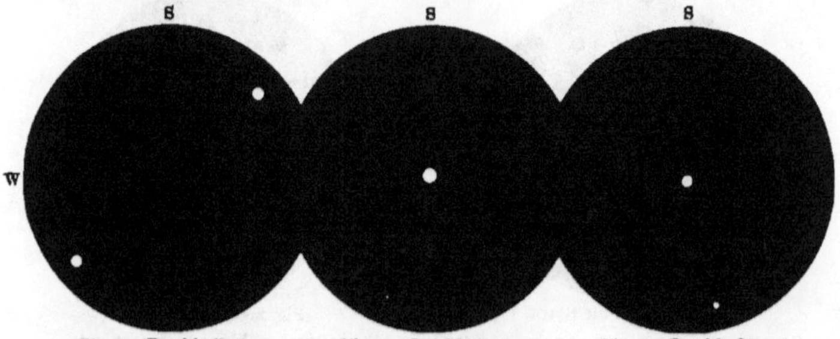

Fig. 3.—Double Star ν   Fig. 4.—Double Star ο   Fig. 5.—Double Star ψ

ψ (*Psi*)—Double; magnitudes 4.8 and 6.0; distance, 31″; yellow and lilac; easy pair.

40—Double; magnitudes 4.5 and 6.0; distance, 20″; very easy.

η (*Eta*)—Double; magnitudes 5.5 and 10; distance, 4″.7; difficult pair.

17—Double; forms a triple with 16; magnitudes 6-6 and 6.5; distances, 4″ and 90″.

ε (*Epsilon*)—Double; magnitudes 4.4 and 8; distance, 2″.9; gold-yellow and azure; difficult pair.

μ (*Mu*)—Binary; magnitudes 5.0 and 5.0; distance, in 1884, 2″.46; in 1781 it was 4″.35; revolution, calculated by M. Berberich, 648 years (Revue d'Ast., Nov., 1885; page 413).

σ (*Sigma*)—One of the nearest stars to us; parallax by Brunnow in 1868 and 1870: average, 0″.25; 838,000 times the distance from the sun to us, or about 77 trillions of miles; it takes the light a little over 13 years to reach us (Revue d'Ast.,1889; page 450).

Σ 1516—Binary; magnitudes 7 and 12; distance, 7″. Mr. L. de Ball has found for the parallax 0″.104±0″.008, which represents 1,983,318 times the distance of the sun from the earth; the light takes 31 years to reach us (Revue d'Ast., Dec. 1887; page 461). It is about 4° N. by E. of λ (*lambda*).

39—Double; magnitudes 5.0 and 7.7; distance, 3″.1; an instrument with good power shows it triple.

Σ 2398 is also one of the nearest stars to us, the average parallax being 0″.33, representing only 55 trillions of miles; the light takes 9 years and 4 months to reach us; it is a double of 8.2 and 9.2 magnitudes, a little west of o (*omicron*). Out of twenty-three stars which gave the best known parallaxes in 1889, this constellation has three of them, σ (*sigma*) only being visible to the naked eye, which proves that the brightest stars are not always the nearest.

17,415 Œltzen is also one of the nearest stars to us; the average parallax being 0″.20.

H. IV, 37—Nebula, the first one examined with the spectroscope, contains nitrogen and hydrogen; before it was doubtful that nebulæ could be in a gaseous state, but the observations by Higgins, in 1861, decided the question; it is near the pole of the ecliptic (see about one-third the distance between ω (*omega*) and 39).

Fig. 6.—Nebula H. IV, 37.

Fig. 7.—Nebula H. IV, 37, in Lick Observatory Telescope.

Fig. 6 represents this nebula as seen with common telescopes. Fig. 7 represents the same nebula as seen by Messrs. Holden and Schœrberle with the large equatorial of Lick Observatory. Attempts have been made several times to find the distance of this nebula.

AUTHORS.	PARALLAX.
Brunnow, in 1871–72	0″.047 ±0″.030
Oudemans	0″.085 ±0″.028
Bredichin, 1876	−0″.005

The last parallax being negative, the distance is very uncertain.

CEPHEUS.

Cepheus, king of Ethiopia and one of the Argonauts, was the husband of Cassiopea and father of Andromeda. This constellation is one of the forty-eight constellations of the ancients and appeared in Eudoxus' astronomical sphere; it is always visible in our latitude.

DESIGNA-TION.	MAGNI-TUDE.	POSITION		DESIGNA-TION.	MAGNI-TUDE.	POSITION	
		R. A. 1880	DECL.			R. A. 1880	DECL.
		h. m.	° ′			h. m.	° ′
α dbl.	2.6	21.16	+62. 5	ρ	6.0	22.29	+78.12
β dbl.	3.4	21.27	70. 2	43 Hev.	4.7	0.51	85.36
γ	3.3	23.34	76.58	51 Hev.	5.5 org.	6.44	87.14
δ dbl.	4.V. org.	22.25	57.48	R	5.V.	20.16	88.46
ε	4.7	22.11	56.27	S	8.V. red	21.37	78. 5
ζ	3.9 red	22. 7	57.37	U	7.0 V.	0.52	81.14
η dbl.	3.9	20.43	61 22	Σ 2843 dbl.	7.0	21.49	65.11
θ	4.4	20.27	62.35	Σ 2840 dbl.	6.5	21.48	55.15
ι	4.0	22.45	65.34	Σ 2895 dbl.	6.5	22.11	72.45
κ bin.	4.5	20.14	77.21	20	6.0 org.	22. 1	62.12
λ	5.8	22. 7	53.51	*	6.0 red	21.30	53.11
μ	4.V. red	21.40	58.14	*	7.5 red	21.10	59.37
ν	5.0	22. 0	59.14	*	5.7 org.	21.53	63. 3
ξ dbl.	5.0	22. 0	64. 2	*	5.9 org.	22. 0	62.31
ο dbl.	5.4	23.13	67.27	*	6.0 org.	22. 1	62.12
π bin.	5.0	23. 4	74.44	*	6.0 org.	22.34	56.10

NOTES.

α (*Alpha*) *Alderamin*—Mr. Pritchard, in 1889, found a parallax of this star=0″.06±0″.02.
δ (*Delta*)—Double variable; magnitudes 3.7 to 4.9 and 7.0; distance, 41″; orange and blue; nice pair; very easy; it varies in the short period of 5 days, 6 hours and 42 minutes. A nearer faint companion discovered by Mr. Burnham; distance, 19″.
β (*Beta*) *Alphirk*—Double; magnitudes 3, 4 and 8; distance, 14″; white and blue; nice pair; not difficult.
γ (*Gamma*)—Is also called Errai.

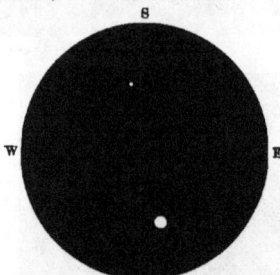

Fig. 8.—Double Star δ

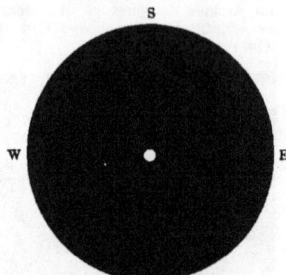

Fig. 9.—Double Star β

κ (*Kappa*)—Binary; magnitudes 4.5 and 8.5; distance, 7″.4; delicate pair.
ξ (*Zi*)—Double; magnitudes 5.0 and 7.6; distance, 6″.6; nice pair. Near to it there is another double star of the 7th magnitude.

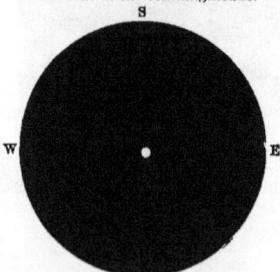

Fig. 10.—Double Star κ

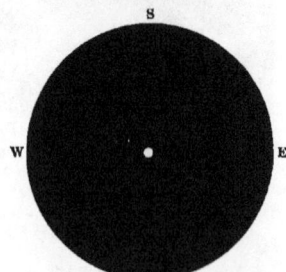

Fig. 11.—Double Star ξ

μ (*Mu*)—Variable; magnitude 4 to 6; of garnet color, called by W. Herschel "*Garnet Sidus;*"
   it is the star of the darkest red, visible to the naked eye. Mr. Burnham discovered a
   faint companion; distance, 19".
ο (*Omicron*)—Double variable; magnitude 5.4 to 7 and 8; distance, 2".5; yellow and blue;
   beautiful contrast.
U.—Varies from 7.5 to 9.2 magnitude in 2 days 11 hours and 50 minutes; it is one of the
   shortest in time of variation; it is N. by W. of γ (*gamma*).
R.—Varies from 6.0 to 10th magnitude in 365 days; it is between α (*alpha*) and δ (*delta*) of
   Ursa Minor.

CAMELOPARDALUS.

This constellation, also called the Camelopard or Giraffe, was introduced by Hevelius,
in 1690; as it was formed about eighty years after Bayer's atlas was published no Greek letters
appear. Our catalogue contains the designation of Flamsteed and Piazzi catalogues.
This constellation, being circumpolar, is visible from our latitude the entire year.

DESIGNA-TION.	MAGNI-TUDE.	POSITION		DESIGNA-TION.	MAGNI-TUDE.	POSITION	
		R. A. 1880	DECL.			R. A. 1880	DECL.
		h. m.	° ′			h. m.	° ′
9	4.6	4.42	+66. 6	P. III, 121	5.5	3.35	+65. 8
10 dbl.	4.2	4.53	60.15	P. IV, 7	5.5	4. 5	53.19
P. III, 111	4.3	3.32	70.58	P. IV, 269 dbl.	5.0	4.54	79. 6
P. III, 51	4.7	3.16	59.32	11 dbl.	5.5	4.56	58.50
P. V, 335	4.9	6. 1	69.22	42	5.5	6.38	67.42
P. VI, 201	4.9	6.35	77.15	43	5.6	6.39	69.
P. XII, 230 dbl.	5.0	12.52	84. 4	P. X, 22	5.5	10. 9	83.10
7 dbl.	5.0	4.47	53.32	R	8.V. org.	14.27	
P. III, 7	5.0	3. 6	65.13	*	6.6 red	3.32	62.15
P. III, 54	5.0	3.18	58.18	*	6.0 org.	3.38	65. 9
P. III, 57	5.2	3.18	54.77	*	5.8 org.	3.47	60.45
1042 Radcl.	5.3	3.35	70.30	*	7.0 org.	4.39	67.57
1 dbl.	5.4	4.27	53.39				

NOTES.

P. IV, 269—Double; magnitudes 5.0 and 8; distance, 19"; star in rapid motion; the distance
   was 37" in 1825, 20" in 1877, and if they continue at the same rate the components will
   be at the nearest point, only 9" apart, in 1932 (Flammarion, " Les Etoiles," page 46).

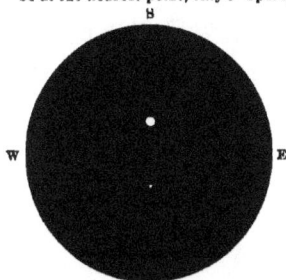

Fig. 12.—Double Star 269.

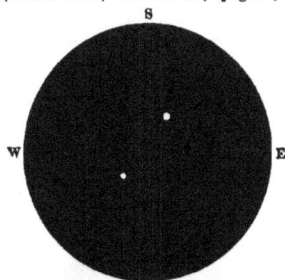

Fig. 13.—Double Star 230.

P. XII, 230—Double; magnitudes 5.8 and 6.4; distance, 22"; very easy pair.
7—Double; magnitudes 4.0 and 11.5; distance, 26"; the companion is a dark ashy color (gray).
11—Double; magnitudes 5.6 and 6.2; distance, 181"; bluish and orange; very easy—an opera
   glass will separate them.

CASSIOPEA.

Cassiopea, or "The Lady in the Chair," is the queen of Ethiopia, wife of Cepheus, who
boasted that her beauty was fairer than the Nereids; those nymphs, to punish the pretentious
queen, prayed Neptune to avenge them; the God of the waters sent a sea monster to ravage
the shores of her kingdom. From another version it was of her daughter's beauty that the
queen boasted. This constellation never sets in our latitude and can be seen every day.

DESIGNA-TION.	MAGNI-TUDE.	R. A. 1880 h. m.	DECL. ° '	DESIGNA-TION.	MAGNI-TUDE.	R. A. 1880 h. m.	DECL. ° '
α dbl.	2.2 V. red	0.34	+55.53	χ	5.7	1.26	+58.37
β dbl.	2.4	0. 3	58.29	ψ quad.	4.5	1.17	67.30
γ dbl.	2.3	0.49	60. 4	ω	5.8	1.46	68. 9
δ	2.8	1.18	59.37				
ε	3.5	1.46	63. 5	48 A bin.	4.7	1.51	70.18
ζ	4.0	0.30	53.22	50	4.2	1.52	71.49
η bin.	4.1	0.42	57.11	P. II, 227	5.0	2.58	73.56
θ	4.4	1. 4	54.31	1	5.3	22.57	56.28
ι trin.	4.5	2.19	66.51	Σ 3062 bin.	5.0	0. 0	57.29
κ	4.5	0.26	62.16	P.XXIII,101 mtp	5.0	23.25	57.44
λ dbl.	5.1	0.25	53.52	4 qdl.	6.0	23.20	61.37
μ dbl.	6.0	1. 0	54.20	Σ 3033 dbl.	6.5	23.57	65.24
ν dbl.	5.6	0.42	50.18	R	6.V. red	23.52	50.43
ξ	5.6	0 35	49.50	S	7.V. red	1.11	71.59
o dbl.	5.2	0.38	47.36	T	7.V. red	0.17	55. 8
π	5.2	0.37	46.22	*	6.5 red	2.47	63.50
ρ	5.3 red	23.48	56.49	*	6.1 org.	2.28	65.14
σ dbl.	5.3	23.53	55. 5	**	7.9 red & blue	23.55	59.41
τ	5.5	23.41	57.59	3077 Bradley	6.5	23. 7	56.30
υ	5.4	0.48	58.19	H. VI, 30	cl.	23.51	56. 3
φ dbl.	5.5	1.12	57.36	Temp. of 1572		0.18	63.27

NOTES.

Mr. Pritchard, in 1888, obtained a parallax for α (alpha)=0″.071±0″.040; and for β (beta) a parallax =0″.162±0″.052. α (alpha) is also called Schedar, and β (beta) Chaph.

γ (Gamma)—Examined with the spectroscope has a double spectrum, similar to the one of the temporary of 1866 (see Corona Borealis notes). It contains some incandescent hydrogen; it has been burning for more than 2,000 years and the fire seems to be as fierce as ever. Mr. Pritchard found, in 1888, for the parallax of this star 0″.007±0″.042; too uncertain to calculate its distance. Mr. Burnham discovered a close star at 2″.2 distance in 1888 with the 36-inch telescope of Lick Observatory.

μ (Mu)—Has a rapid, proper motion; 4″.43 per year; one degree in 812 years; apparent diameter of the moon in 420 years; will be near φ (phi) Persei in 6,000 years if it keeps the same minimum velocity of 200,000 yards per second. It was "parallaxed" by Struve in 1855, by Schweizer in 1863 and by Pritchard in 1888, and the average 0″.060 would put the distance at 3,438,000 times the distance of the sun from the earth or 318 trillions of miles, and the light would have to travel 54 years before reaching us (Revue d'Ast., 1889; page 450).

η (Eta)—Binary; magnitudes 4.2 and 7; distance, in 1880, 5″.3; revolution, about 167 years. It was noted yellow and lilac by Flammarion; red and green in 1821, by J. Herschel; yellow and purple in 1832, by Struve; yellow and blue in 1841, by Dawes; yellow and red in 1851, by Secchi; the companion seems to change color. This star showed a parallax of 0″.154, measured by Struve in 1855, and one of 0″.170, measured by Schweizer in 1866. In taking 0″.16 as an average it would be at 1,270,000 times the distance of the earth to the sun or 118 trillions of miles, and the light would take over 20 years to reach us (Revue d'Ast., 1889; page 446).

ι (Iota)—Trinary; magnitudes 4.5-7.0 and 8.4; distances, 2″ and 7″.6; yellow, lilac and purple; remarkable system.

ψ (Psi)—Quadruple or double-double; magnitudes 4.5 and 13; 9 and 10; the main stars are 20″ apart, and form two difficult pairs at 3″ distance each. The companion of the largest star was discovered by Mr. Burnham in 1889.

3077 Bradley—Parallax by Brunnow in 1871, Backlund 1881 and Gylden 1881. Average ÷0″.19.

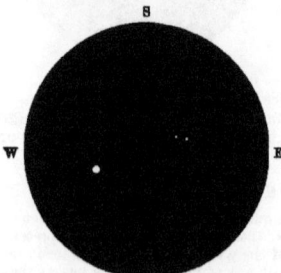

Fig. 14.—Quadruple Star ψ

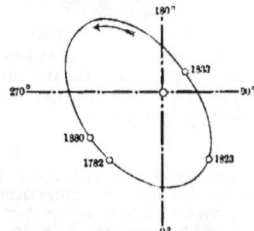

Fig. 15.—Orbit of Σ 3062.

P. XXIII, 101—Multiple; eight stars have been measured in the group, which include two close pairs 1".4 and 1".5 apart, but it requires a good power to see them all.

σ (*Sigma*)—Double; magnitudes 5.3 and 8; distance, 3"; delicate pair.

Σ 3062—Binary; magnitudes 6.9 and 7.5; distance, 1".4; rapid orbital motion; revolution, about 104 years.

T (1572)—Temporary; very remarkable; observed by Tycho Brahe, it was first visible in daytime, very much to the astonishment of the people; during five months it was brighter than the stars of the 1st magnitude, but gradually decreased to the 2d, 3d, 4th, etc., magnitude, and 17 months afterward disappeared to the naked eye; the telescope was not then invented and it could not be followed any longer. When it first appeared it was as white and as bright as Venus; after it turned yellow, and finally it was seen very red. A similar temporary is reported to have appeared in the same place in 945 and in 1264, which may lead to the conclusion that it was the same star, reviving at intervals of 319 and 308 years, but the position was not positively defined, and if it was the same star it should have appeared again in 1880. A star of the 11th magnitude is near its position.

ζ (*Zeta*)—Prof. Colbert, of Chicago, said that on the 20th of August, 1886, at 9.30 p. m. he saw this star "as appearing to be more prominent than a star in the same region which is rated in the catalogues as half a magnitude brighter, and nearly as bright as the most prominent stars in that region." It lasted only half an hour and then went down to the regular size; the strangest part of it is that at the moment of its greatest brilliancy a ray of light was seen similar to the tail of a small comet going from it in the direction of λ (*lambda*).

H. VI., 30—Nice cluster of small stars—they look like diamond dust (see between β (*beta*) and σ (*sigma*).) (Fig. 16.)

Fig. 16.—Cluster H. VI, 30.

ANDROMEDA.

Andromeda, or "The Chained Lady," was the daughter of Cepheus and Cassiopea; to appease the anger of the Nereids, whom her mother had offended in boasting of her beauty, she was attached to a rock to be killed by the sea monster that Neptune sent to ravage the shores of Ethiopia, but Perseus rescued her and made her his wife.

DESIGNA- TION.	MAGNI- TUDE.	POSITION		DESIGNA- TION.	MAGNI- TUDE.	POSITION	
		R. A. 1880 h. m.	DECL. ° '			R. A. 1880 h. m.	DECL. ° '
α dbl.	2.0	0. 2	+28.26	φ dbl.	4.5	1.36	+50. 5
β dbl.	2.2 red	1. 3	34.59	χ	5.6	1.32	43.48
γ trin.	2.1 org.	1.56	41.45	ψ dbl.	5.7	23.40	45.45
δ dbl.	3.3 yel.	0.33	30.13	ω	4.7	1.20	44.48
ε	4.3	0.32	28.40				
ζ	4.3	0.41	23.37	A	6.0	1.23	46.24
η	4.4	0.51	22.47	b	5.5	2. 5	43.40
θ	5.4	0.11	38. 1	c	6.0	2.11	46.50
ι	4.5	23.32	42.36	53	4.8	1.32	39.50
κ trip.	4.5	23.35	43.40	3 dbl.	5.5	22.53	49.23
λ	4.4 yel.	23.32	45.49	7	5.4	23. 7	48.45
μ	4.3	0.50	37.51	8	5.0 red	23.12	48.22
ν	4.5	0.43	40.26	41	5.4	1. 1	43.19
ξ	5.0	1.15	44.54	55 dbl.	6.0	1.46	40.10
ο	4.0	22.57	41.41	56 dbl.	5.5	1.49	36.40
π dbl.	4.3	0.30	33. 4	59 dbl.	6.5	2. 4	33.29
ρ	6.0	0.15	37.18	R	6.V. org.	0.18	37.55
σ	4.7	0.12	36. 7	*	8.2 very red	0.14	44. 3
τ	4.6	1.34	39.58	34 Groomb.	8.0	0.11	43.20
υ	5.5	1.30	40.49	M. 31	neb.	0.36	40.37

NOTES.

The principal stars are named: α (*alpha*), Alpherat; β (*beta*), Mirach; and γ (*gamma*), Almach.

γ (*Gamma*)—Trinary; magnitudes 2.2-5.5 and 6.5; distances, 10″ and 0″.5; orange, green and blue; splendid system. The second and third form a system in orbital motion, period unknown. Mr. Burnham said that the double companion was single in all telescopes for some years, and the distance now (Dec. 1891) is less than 0″.1.

π (*Pi*)—Triple; double with small instrument; magnitudes 4.4 and 9; distance, 36″; easy pair.

56—Double; magnitudes 6 and 6; distance, 2′ 56″; very easy pair—an opera glass will separate them.

36—Binary; magnitudes 6 and 7; distance, 1″.3; close pair; orange and yellow; revolution, 137 years (see between β (*beta*), η (*eta*) and ζ (*zeta*).

R.—Variable; sometimes visible to the naked eye; it varies from the 6th to the 13th magnitude in 405 days.

M. 31—Beautiful nebula, visible with an opera glass—even with the naked eye. Halley, who observed it, says: "*The spot is nothing else than the light coming from an extraordinary great space in the ether, through which a lucid medium is diffused which shines with its own proper luster.*" Bond gives it 4° in length and 2°30″ in width, which represents a minimum of 1,381 billions of miles, or 300 times larger than the solar system. Fig. 18 is an exact reproduction of a photograph of this nebula taken by Mr. Isaac Roberts, December 29, 1888, with a 20-inch telescope. It took four hours to obtain this result.

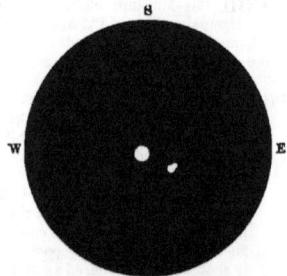

Fig. 17.—Trinary γ

Fig. 18.—Nebula M. 31.
(From a photograph taken by Mr. Isaac Roberts.)

The spectral analysis indicated that this nebula is entirely gaseous; it is nearly circular, and its elliptic aspect is due to its great obliquity. Another photograph, taken in 1891 by Mr. Roberts, shows some spiral curves with several condensations, indicating a world in formation, conformable to the theory of Mr. Faye (Revue d'Ast., 1891; pages 441-445).

T. (1885)—In 1885 a temporary appeared near the center of the nebula. We give below the table of the variations of magnitude observed:

1885	MAG.	1885	MAG.	1885	MAG.
Aug. 17	9	Sept. 2	6½	Sept. 20	9
Aug. 19	9	Sept. 3	6½	Sept. 27	10
Aug. 22	9	Sept. 4	7	Oct. 1	10½
Aug. 30	6½	Sept. 7	7½	Oct. 8	11
Aug. 31	6½	Sept. 11	8	Oct. 15	11½
Sept. 1	6¼	Sept. 15	8½		

It was most likely a conflagration (Revue d'Ast., 1885; pages 361, 403, 408).

TRIANGULUM.

This constellation was called by the Greeks the "Deltoton;" some astronomers called it Triangula (the Triangles); it was already one of the constellations in Eudoxus' time (in the 4th century B. C.).

In our map of the constellations at the end of this Handbook appears a *fly* above Aries; this little constellation (*Musca*) was introduced by Bartschius, in 1624, but did not last long. The B. A. Catalogue does not recognize it.

DESIGNA-TION.	MAGNI-TUDE.	POSITION		DESIGNA-TION.	MAGNI-TUDE.	POSITION	
		R. A. 1880 h. m.	DECL. ° '			R. A. 1880 h. m.	DECL. ° '
a dbl.	4.0	1.46	+29. 0	ε dbl.	5.8	1.56	+32.42
β	3.2	2. 2	34.25	6 dbl.	5.8	2. 5	29.45
γ	4.2	2.11	33.23	7	6.0	2. 9	32.48
δ	5.5	2.10	33.42	M. 33	neb.	1.27	30. 2

NOTES.

6—Double; magnitudes 5.5 and 6.5; distance, 3''.7; gold-yellow and bluish-green; very nice pair.

M. 33—Nebula, large but not well defined; find it when there is no moonlight, between a (*alpha*) of Triangulum and β (*beta*) of Andromeda.

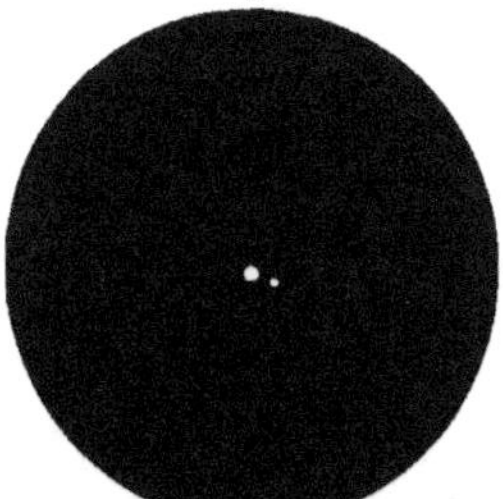

Fig. 19.—Double Star 6.

LACERTA.

This little constellation has very little interest, and Hevelius said that he noticed ten very bright little stars between Andromeda and Cygnus, and he inserted a lizard because he could not put anything else more appropriate.

Bode, in 1798, introduced between Lacerta, Cassiopea and Andromeda a constellation in honor of the King of Prussia, Frederick; it was composed of stars taken out of the constellation Cassiopea, but the B. A. Catalogue does not recognize it. *Honores Frederici* may be seen in our map of the constellations at the end of this Handbook.

Lacerta itself was a constellation formed by Augustin Royer, in 1679, in honor of Louis XIV of France, but the name given by Hevelius, in 1690, has been adopted.

DESIGNA-TION.	MAGNI-TUDE.	POSITION		DESIGNA-TION.	MAGNI-TUDE.	POSITION	
		R. A. 1880 h. m.	DECL. ° '			R. A. 1880 h. m	DECL. ° '
7 Fl. a	4.2	22.47	+49.40	6	5.2	22.25	+42.30
3 β	4.7 org.	22.19	51.38	10 dbl.	5.2	22.34	38.25
1 dbl.	4.8 org.	22.11	37. 9	11	5.5 yel.	22.35	43.38
2 dbl.	4.8	22.16	45.56	15 dbl.	5.5 org.	22.47	42.41
4	5.0 org.	22.20	48.52	P. XXII, 36	5.3 red	22. 9	39. 7
5	5.0 red	22.24	47. 5				

NOTES.

4—Is an orange star; near by is a blue star; nice field, rich in small stars.

8 Fl.—Is a quadruple star near 10.

PERSEUS.

Perseus, also called "The Champion," is the hero who, after hearing of the dangers of Andromeda, jumped on Pegasus and arrived in time to save her life by presenting to the sea monster the head of Medusa, which had the power of petrifying everything and everybody. It is also one of the oldest constellations, noted by Eudoxus.

DESIGNA-TION.	MAGNI-TUDE.	POSITION R. A. 1880 h. m.	POSITION DECL. ° ′	DESIGNA-TION.	MAGNI-TUDE.	POSITION R. A. 1880 h. m.	POSITION DECL. ° ′
α dbl.	2.2	3.16	+49.26	58 e, trip.	4.6 org	4.28	+41. 1
β dbl.	2.V. red	3. 0	40.30	52 f	5.0	4. 7	40.11
γ dbl.	3.0	2.56	53. 2	4 g	5.6	1.54	53.54
δ dbl.	3.5	3.34	47.24	h	cum.	2. 5	50.31
ε dbl.	3.3 V.	3.50	39.40	9 i	5.7	2.14	55.17
ζ qdl.	3.0	3.47	31.32	k	5.2	2.56	56.14
η dbl.	4.2 red	2.42	55.24	l	5.5	3.13	42.55
θ trip.	4.4	2.36	48.43	57 m, dbl.	6.5	4.25	42.47
ι	4.3	3. 0	49. 9	42 n	6.6	3.42	32.42
κ	4.4	3. 1	44.24	40 o, dbl.	5.7	3.35	33.34
λ	4.6	3.58	50. 2	16	4.5	2.43	37.49
μ dbl.	4.5	4. 5	48. 6	17	5.0	2.44	34.34
ν	4.1	3.37	42.12	21	5.2	2.50	31.26
ξ	4.3	3.51	35.27	995 B. A. C.	5.2	2.43	50.29
ο dbl.	4.3	3.35	31.55	29-31	5.4	3.10	49.45
π	5.1	2.51	39.10	P. III, 23	5.4	3.10	33.11
ρ	3.V. red	2.57	38.22	24	5.5	2.51	34.42
σ	4.8 org.	3.22	47.35	12 trip.	5.5	2.34	39.40
τ	4.3	2.46	52.16	P. II, 220 dbl.	5.8	2.53	51.52
υ	3.9	1.30	48. 1	Σ 563, dbl.	7.5	4.28	40.51
φ	4.0	1.36	50. 5	R	8.V. org.	3.22	35.16
χ trip.	cum.	2.10	56.58	S	8.V. org.	2.14	58. 2
ψ	4.8	3.28	47.47	H. VI, 33	cl.	2.11	56.36
ω	5.0 red	3. 3	39. 9	II. VI, 34	cl.	2.14	56.33
				M. 34	cl.	2.34	42.16
43 A, dbl.	5.6	3.48	50.21	neb. and * red.		2.36	31.55
b	5.1	4. 9	50. 0	*	7.5 red	2.43	57.50
48 c	4.4	4. 0	47.23	*	7.5 red	3.21	54.58
43 d	5.3	4.13	46.12				

NOTES.

α (Alpha) Mirfak—Is a double star.

β (Beta) Algol—Very remarkable variable; visible to the naked eye; varies from 2.3 to 4.3 in 2 days 20 hours 48 minutes and 53 seconds; the minimum lasts only 6 minutes and the maximum about 36 hours, consequently the variation takes only about 9 hours. Is it due to the rotation of Algol, which would have a dark continent? or is it due to the revolution of a planet which would partially eclipse it? From spectroscopical observations done at Greenwich the latter is probably true, but it is not yet proved (Revue d'Ast., Nov., 1887; page 428).

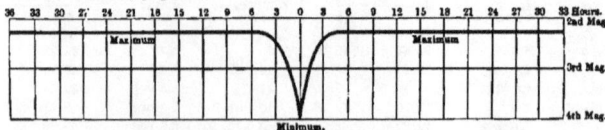

Fig. 20.—Diagram showing the variations of Algol in 69 hours.

ρ (Rho)—Is also a variable; from 3.4 to 4.2; period not yet known.

η (Eta)—Double; magnitudes 4.2 and 8.5; distance, 28″; yellow and blue; this fine pair has five little stars around it.

ε (Epsilon)—Double; magnitudes 3.3 and 8.5; distance 9″; greenish-white and lilac.

θ (*Theta*)—Triple; magnitudes 4.4-10 and 10; distances, 15″ and 68″.

ζ (*Zeta*)—Quadruple; magnitudes 3-10-12 and 11; distances, 13″, 83″ and 121″; very difficult.

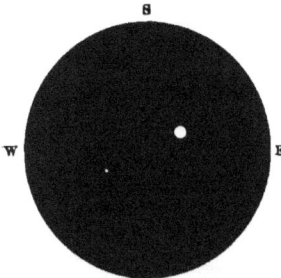

Fig. 21.—Double Star η

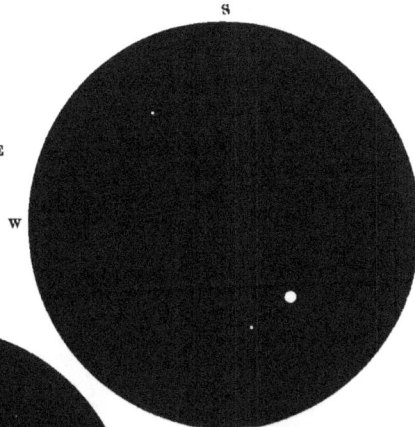

Fig. 22.—Triple Star θ

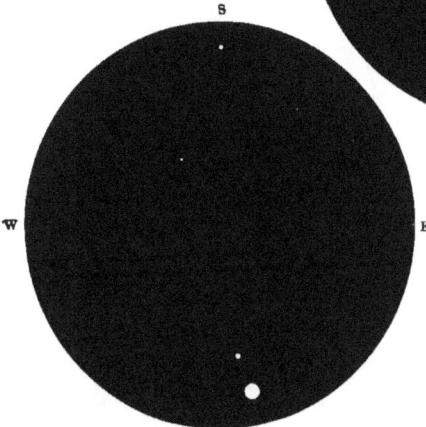

Fig. 23.—Quadruple Star ζ   Scale—60″ 1 inch.

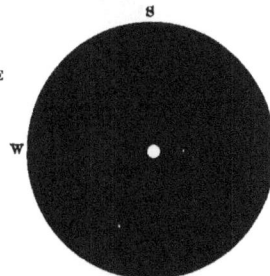

Fig. 24.—Double Star ε

P. II. 220 Double; magnitudes 6 and 8; distance, 12″; very easy pair; the companion is sapphire.

Σ 563 Double; magnitude 7.5 and 9; distance, 12″; delicate pair.

H. IV, 33 and 34—Are two clusters visible to the naked eye, very close to each other and composed of several hundreds of stars; with a small power and large field it is a very nice sight; it is marked in our planisphere N. by W. of η (*eta*).

M. 34—Is also a nice cluster of stars, very easy with small instruments; it is between Algol and Almach; it was resolved into small stars by Messier himself. This constellation is rich in nebulæ, more or less easy with common telescopes.

Medusa's Head (Caput Medusæ) is the name given to a cluster of stars in this constellation, of which Algol is a part.

Fig. 25.—Cluster Messier 34.

URSA MAJOR.

According to Greek mythology "The Great Bear" is nothing else than the nymph Callisto, who was beloved by Jupiter, and became the mother of Arcas; according to Ovid, the jealous Juno to avenge herself changed Callisto into a bear, and one day when Arcas was hunting he did not recognize her and came very near killing his mother; to avoid this parricide Jupiter carried them both among the stars.

It is mentioned in Job (xxxviii, 31) and by Homer, and is perhaps the oldest of the constellations. The stars of this constellation never set in our latitude, and a (alpha) and β (beta), which are called "The Pointers," serve to find the north point, as the Pole Star is in the direction of these stars at about five times the distance that separates them.

This constellation is commonly named "The Dipper;" the French people call it "Chariot of David;" the Chinese, Ti-tche ("Chariot of the Sovereign"). The four stars a (alpha), β (beta), γ (gamma) and δ (delta) are called by the Arabs "The Coffin."

DESIGNA-TION.	MAGNI-TUDE.	POSITION		DESIGNA-TION.	MAGNI-TUDE.	POSITION	
		R. A. h. m.	1880 DECL. ° ′			R. A. h. m.	1880 DECL. ° ′
α dbl.	2.4 V. yel.	10.56	+62.24	b	5.5	8.43	+62.24
β dbl.	2.8	10.55	57.02	c	5.5	9. 5	61.55
γ dbl.	2.7	11.48	54.22	d	5.2	9.24	70.22
δ dbl.	3.7	12. 7	57.41	e	5.0	9. 7	54.32
ε bin.	2.2	12.40	56.37	f	5.2	9. 0	52. 5
ζ (Mizar) dbl.	2.4	13.19	55.33	g	5.0	13.20	55.37
η	2.1	13.43	49.55	23 h, dbl.	4.2	9.22	63.36
θ dbl.	3.3	9.25	52.13	10	4.5	8.52	42.15
ι dbl.	3.4	8.51	48.31	P. VIII, 245	5.0	8.59	38.56
κ	3.4	8.53	47.38	26	5.4	9.26	52.35
λ	3.3	10.10	43.31	P. X, 42	5.0	10.14	66.10
μ	3.2 red	10.15	42.06	38	5.2	10.34	66.21
ν dbl.	3.3 red	11.12	33.45	P. X, 135	5.3	10.36	46.50
ξ bin.	3.6	11.12	32.12	47	5.3	10.53	41. 4
ο	3.8	8.20	61. 7	49	5.5	10.54	39.51
π	5.0	8.29	64.45	55	5.5	11.12	38.51
ρ bin. (?)	5.2	8 52	68. 5	57 dbl.	5.9	11.23	40. 0
σ dbl.	5.3	9. 0	67.37	83	5.V. org.	13.36	55.18
τ dbl.	5.5	9. 3	63.59	1830 Groomb.	6.7	11.46	38.35
υ dbl.	4.8	9.42	59.37	21185 Lal.	7.5	10.56	36.53
φ bin.	5.0	9.44	54.38	21258 Lal.	8.5	11. 0	48. 7
χ	4.0 red	11.40	48.26	R	7.V. red	10.36	66.24
ψ	3.2 yel.	11. 3	45. 9	S	8.V. org.	12.39	61.45
ω	5.0	10.47	43.50	T	7.V. org.	12.31	60. 9
				P. X, 126	7.V. org.	10.34	69.42
A	5.5	8.24	65.34				

NOTES.

The first seven stars of this constellation have been named by the Arabs as follows: a (alpha), Dubhe; β (beta), Merak; γ (gamma), Phegda; δ (delta), Megrez; ϵ (epsilon), Alioth; ζ (zeta), Mizar; and η (eta), Benetnash or Alkaid.

a (Alpha) has a companion discovered by Mr. Burnham in 1889; distance $0''.9$.

Mizar and Alcor are visible to the naked eye; distance $11' \, 48''$.

Fig. 26 represents Mizar and Alcor as seen in the field of an ordinary telescope; the two stars of the West represent the double star Mizar; the N. E. star is Alcor; the little stars which appear on the diagram are difficult to be seen with small instruments. Mizar was observed as a double star by Riccioli, in 1650; by Gottfried Kirch at the end of the 17th century. The observations of the components taken at different times indicate that the relative position of the two stars varied only a few degrees in 125 years; the revolution around their center of gravity exceeds 18,000 or 20,000 years. (Flammarion, Les Etoiles, page 107.)

Scale—6′ 30″—1 inch.

S

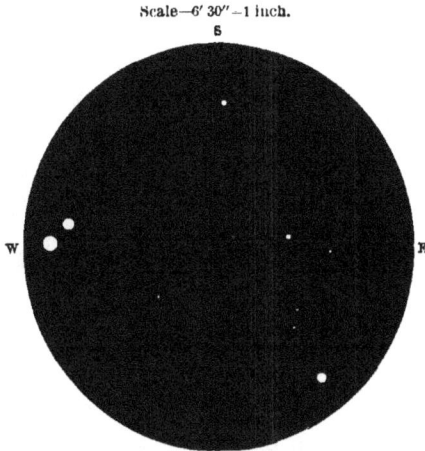

Fig. 26.—The Double Star Mizar and Alcor, in the field of any ordinary telescope.

ζ *(Zeta) Mizar*—Double; magnitudes 2.4 and 4.0; distance, 14″.5; splendid pair, very bright. It is the first star observed as double in the telescope.

ξ *(Zi)*—Binary; magnitudes 3.6 and 5; distance, in 1880, 1″.7; rapid orbital motion; revolution, 60 years and 7 months. This system is the first one which had its period calculated by Savary, in 1828, he giving for the revolution 58 years.

ι *(Iota) Talitha*—Is also double.

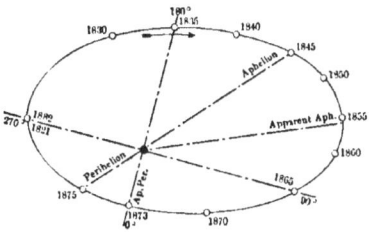

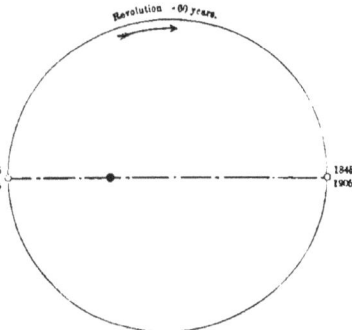

Fig. 27.—Apparent Orbit of ξ in Ursa Major. Fig. 28.—Real Orbit of ξ

ν *(Nu)*—Double; magnitudes 3.3 and 10; distance, 7″; orange and blue.

σ *(Sigma)*—Double; suspected binary; magnitudes 5.3 and 9; distance, 2″.6; in 1780 it was at distance 8″; there are two σ *(sigma)*, the double is σ².

57—Double; magnitudes 5.9 and 8; distance, 5″.5; nice pair; the companion is violet.

φ *(Phi)*—Binary; magnitudes 5 and 5.5; distance, 0″.24 in 1892; close pair; time of orbital revolution, 115 years.

OΣ 234—Binary; magnitudes 7 and 7.8; distance, 0″.2 in 1892; revolution, 68 years.

OΣ 235—Binary; magnitudes 6 and 7; distance, 1″ in 1892; revolution, 94 years.

ι *(Iota)*—Peters, in 1842-43, obtained for the parallax of this star, 0″.068 ±0″.047.

10—Mr. Belopsky, in 1888, obtained for the parallax of this star, 0″.20 ±0″.011.

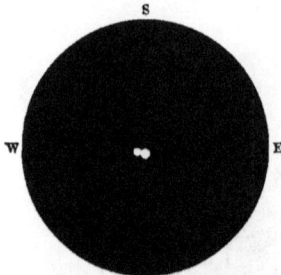

Fig. 29.—Binary ξ in 1880.

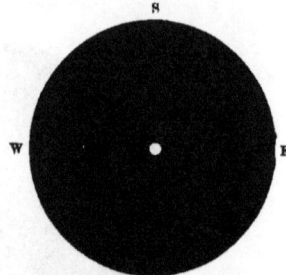

Fig. 30.—Double Star 23 h.

23 h—Double; magnitudes 4.2 and 9; distance, 22″. This pair is stationary since 1781.

1830—Near 57 is the star of the greatest motion, 7″.3 S. E. per year; if it was turning around us it would take 180,000 years to complete its revolution; minimum velocity, 300,000 yards per second; it will be in Coma Berenices 6,000 years from now. The parallax of this remarkable star has been determined several times, and in adopting the average 0″.045 it would represent 4,583,000 times the distance of the sun from the earth, or 500 trillions of miles, and the light would have to travel 72 years to reach us.

AUTHORS.	PARALLAX.	AUTHORS.	PARALLAX.
Peters, 1842–43	0″.226 ± 0″.141	Johnson, 1850–53	0″.033 ± 0″.028
Schlüter and Wichman, 1842–43	0″.182 ± 0″.018	Auwers, 1874	0″.023 ± 0″.033
Schlüter and Wichman, 1850–52	0″.141 ± 0″.013	Brunnow, 1870–71	0″.090 ± 0″.025
O. Struve	0″.034 ± 0″.029	J. C. Kapteyn, 1891	0″.139 ± 0″.026

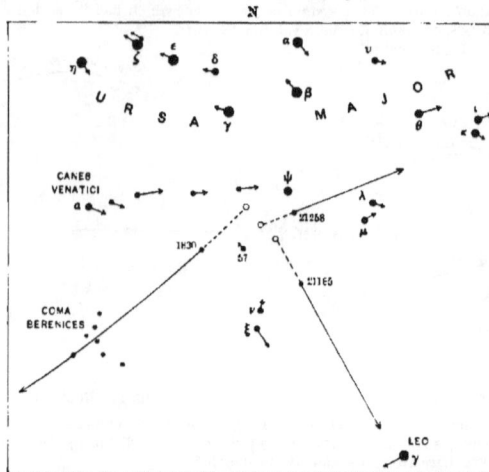

Fig. 31.—Rapid motions of three stars in Ursa Major.

21185–21258—Are also in very rapid motion: in calculating their position it has been found that the three stars, 1830, 21185 and 21258, were very close together 3,000 years ago, and since then they have been moving in three different directions as a result of a "fantastic explosion." These stars, which are not visible to the naked eye, would be found near the double star 57; the first one, as we have said, goes toward Coma Berenices; the second in the direction of γ (gamma) of Leo Major, and the third toward κ (kappa) (Flammarion, Les Etoiles, page 115).

The diagram (Fig. 31) represents the motions of the above three stars; the black circles indicate their position in 1880, the small circles their position 3,000 years ago, and the end of the arrows their position 10,000 years from now.

Mr. Winnecke, in 1858, gave for the parallax of the
star 21185.....................................0″.501 ± 0″.011
Mr. J. C. Kapteyn, in 1891.......................0″.428 ± 0″.030
Mr. Auwers gave for the parallax of the star 21258..0″.262 ± 0″.011
Mr. Krueger.....................................0″.260 ± 0″.020
Mr. J. C. Kapteyn, in 1891.....................0″.168 ± 0″.027

M. 97—This curious nebula, visible only in large telescopes, will be found at about 10° S. E. of β (beta). (See Fig. 32.)

Fig. 32.—Nebula M. 97.

LEO MINOR.

This little constellation was introduced by Hevelius about 1660.

DESIGNA-TION.	MAGNI-TUDE.	POSITION R. A. 1880 h. m.	DECL. ° ′	DESIGNA-TION.	MAGNI-TUDE.	POSITION R. A. 1880 h. m.	DECL. ° ′
37 qdl.	4.9	10.32	+32.35	31	4.4	10.21	+37.18
30	4.9	10.19	34.24	21	4.5	10. 0	35. 5
42 dbl.	5.0	10.39	31.20	10	5.0	9.27	36.56
46	4.2	10.47	34.52	R	7.V. org.	9.38	35. 4

NOTES.

This constellation contains nothing remarkable with the exception of R, between 21 and 10, a little south of the line joining these stars, which varies from the 7th to the 11th magnitude in 369 days.

20—Mr. J. C. Kapteyn, in 1891, gave for the parallax of this star, 0″.062 ± 0″.029.

CANES VENATICI.

This constellation was formed by Hevelius, in about 1690, by taking some stars situated between the Great Bear and the Herdsman.

DESIGNA-TION.	MAGNI-TUDE.	POSITION R. A. 1880 h. m.	DECL. ° ′	DESIGNA-TION.	MAGNI-TUDE.	POSITION R. A. 1880 h. m.	DECL. ° ′
12 a, dbl.	2.9	12.50	+38.58	6	5.2	12.20	+39.41
8 β	4.4	12.28	42. 0	P. XII, 29	5.6	12.10	33.44
14	5.0	13. 0	36.27	P. XIII, 27	5.2	13. 9	40.48
15	5.7	13. 4	39.12	2 dbl.	6.5	12.10	41.20
19	6.0	13.10	41.29	23793 Lal.	5.V. org.	12.39	46. 6
20	5.0	13.12	41.12	*	6.0 org.	13.18	37.40
23	6.0	13.15	40.46	M. 51	neb.	13.25	47.50
21	5.2	13.13	50.19	M. 3	cl.	13.37	28.59
24	4.8	13.30	49.38	M. 94	neb.	12.46	41.48
25 bin.	5.2	13.32	36.54				

NOTES.

a (Alpha)—Double; magnitudes 3.2 and 5.7; distance, 20″; gold-yellow and lilac; very nice pair. It was called by Halley " Cor Caroli II " (Charles II's Heart).

2—Double; magnitudes 6 and 9; distance, 11″; gold-yellow and azure; elegant pair.

25—Binary; magnitudes 6 and 7; distance, 1″.0 in 1892; white and blue; time of revolution, 124 years.

M. 51—Beautiful nebula, with two nuclei. The telescope of Lord Rosse in 1845 shows this nebula in nice spiral curves composed of brilliant dust, each part being a sun like ours, and separated by millions and millions of leagues. Diameter equals 6′ (minutes). (Figs. 34 and 35.)

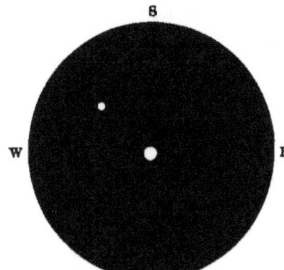

Fig. 33.—Double Star a

Fig. 34.—Nebula M. 51, in Common Telescopes.

Fig. 35.—Nebula Messier 51, in Lord Rosse's Telescope.   Fig. 36.—Cluster M. 3.

M. 3—Rich cluster of 6 to 7 minutes in diameter, containing about 1,000 stars; three small stars in triangular shape seem to inclose it.

COMA BERENICES.

Coma Berenices (Queen Berenice's Hair) is the only constellation of the ancients of which we can give the true history.

Berenice, daughter of King Ptolemy Philadelphus, had just been married to her brother Ptolemy Euergetes, when he was obliged to fight against Seleucus II, King of Syria; Berenice in her grief swore by Venus to sacrifice her beautiful hair if her husband came back victorious; he did, and Berenice, the very day of his return, deposited her hair in the temple of the goddess; the next night it disappeared, stolen, most likely, by a priest; but to console the two lovers the astronomer Conon told them that he saw it in the sky, and that it had been transported there by Venus herself.

It is noted for the first time in the catalogue of Tycho Brahe in 1590; but mention is made of it already by Ptolemy, A. D. 140, by Callimachus, Eratosthenes and Sufi.

DESIGNA- TION.	MAGNI- TUDE.	POSITION		DESIGNA- TION.	MAGNI- TUDE.	POSITION	
		R. A. 1880	DECL.			R. A. 1880	DECL.
		h. m.	° ′			h. m.	° ′
43 β	4.6	13. 6	+28.29	35 trin.	5.7	12.47	+21.56
15 γ	4.9 org.	12.21	28.57	36	5.4 org.	12.53	19. 4
16	5.2	12.21	27.30	37	5.6	12.54	31.25
42 α bin.	5.2	13. 4	18.10	41	5.5	13. 2	28.17
6	5.7	12.10	15.34	7	5.8	12.10	24.39
11	5.5	12.15	18.29	18	6.0	12 23	24.46
12 dbl.	5.4	12.16	26.31	21	6.0	12.25	25.15
14	5.5	12.20	27.56	R	7.V. red	11.58	19.27
23	5.5	12 29	23.17	R	8.V.	12.33	17.10
24 dbl.	5.6	12.29	19. 2	R	7.V.	12.34	17. 8
27	5.8	12.41	17.15	*	8.0 org.	12.52	18.24
31	5.7	12.46	28.12	*	6.0 org.	13.31	25.13

NOTES.

24—Double; magnitudes 5.6 and 7; distance, 21"; orange and blue; beautiful pair; in rapid motion.

35—Trinary; magnitudes 5.7-8 and 8.2; distances, 28" and 1".4; the two companions revolve around each other in a period of 400 to 500 years.

12—Double; magnitudes 5.4 and 8; distances, 66"; very easy pair.

42—Binary; magnitudes 6 and 6; distances, 0".5; very close pair, in orbital motion; revolution only 25 years.

This constellation is very rich in small nebulæ, but it requires good power to see them.

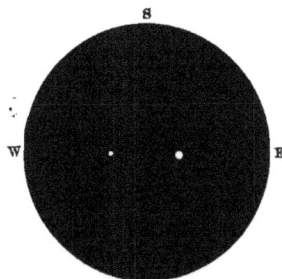

Fig. 37.—Double Star 24.

BOOTES.

This constellation, now called the Herdsman, was also known as Arctophylax, which means "Guardian of the Bear." According to some authors it was Arcas, son of the nymph Callisto; according to others it represents Icarus, the son of Dœdalus. It is one of the forty-eight constellations of the ancients.

DESIGNA-TION.	MAGNI-TUDE.	POSITION R. A. 1880 h. m.	DECL. ° '	DESIGNA-TION.	MAGNI-TUDE.	POSITION R. A. 1880 h. m.	DECL. ° '
α	1.2 org.	14.10	+19.48	A	5.0	14.13	+36. 3
β dbl.	3.3	14.57	40.52	46 b	6.0	15. 3	26.46
γ	3.6	14.27	38.50	45 c	5.7	15. 2	25.21
δ dbl.	3.4	15.11	33.46	12 d	5.7	14. 5	25.40
ε bin.	2.4 yel.	14.40	27.35	6 e	5.8	13.44	21.52
ζ bin. (?)	3.3	14.35	14.15	22 f	6.0	14.21	19.46
η dbl.	3.0	13.49	19. 0	24 g	6.0	14.24	50.23
θ	4.4	14.21	52.24	38 h	6.2	14.45	46.37
ι trip.	4.6	14.12	51.55	44 i, bin.	5.0	15. 0	48. 8
κ dbl.	5.0	14. 9	52.21	47 k	5.9	15. 1	48.36
λ	4.5	14.12	46.38	9	5.5	13.51	28. 6
μ trin.	4.4	15.20	37.48	13 dbl.	5.5 org.	14. 4	50. 1
ν	4.8	15.27	41.15	20	5.5	14.14	16.51
ξ bin.	4.5	14.46	19.37	4559 B. A. C.	5.5	13.34	11.22
ο	4.9	14.40	17.28	P. XIV, 69 dbl.	5.3	14.18	8.58
π bin.	4.3	14.35	16.56	P. XIV, 73	5.5	14.18	6.22
ρ dbl.	4.0 org.	14.27	30.54	31	5.0	14.36	8.41
σ	5.0	14.29	30. 16	34	5.V. org.	14.38	27. 2
τ dbl.	5.0	13.42	18. 3	40	5.8	14.55	39.41
υ	4.8 red	13.44	16.24	39 dbl.	5.6	14.46	49.13
φ	5.3	15.33	40.44	R	6.V. red	14.32	27.15
χ	5.2	15. 9	29.37	S	8.V.	14.19	54.21
ψ	5.0	14.59	27.25				
ω	5.3	14.57	25.29				

NOTES.

α (Alpha) Arcturus—In rapid proper motion; 2".25 S.W. per year. It is the first star observed as being in motion by Halley in 1717; velocity more than 100,000 yards per second; the spectral analysis indicates that it is coming our direction at the rate of about 3,100 miles per minute (Flam., Les Etoiles, page 135). The parallax of Arcturus obtained by Peters in 1842-43, was 0".127+0".073; by Johnson, in 1853-54, was 0".138+0".052; by Elkin, in 1888, was 0".018+0".022. The average, 0".094, would bring the distance of Arcturus at 2,194,100 times the distance from the earth to the sun, or 200 trillions of miles, and the light would take over 34 years and 6 months to reach us (Revue d'Ast., 1889; page 446).

β (*Beta*) is called Nekkar, ε (*epsilon*) Izar, and η (*eta*) Muphrid.

34—Varies from the 4.5 to the 6th magnitude in a period calculated by Schmidt to be 369 days.

ε (*Epsilon*)—Binary; magnitudes 2.4 and 6.5; distance, 2″.9; gold and blue; Struve called it "Pulcherrima" (the finest); orbital revolution over 1,200 years.

π (*Pi*)—Binary; magnitudes 4.3 and 6; distance, 6″; nice and easy pair.

ι (*Iota*)—Double; magnitudes 4.6 and ⁻8; distance, 38″. Triple with large telescopes.

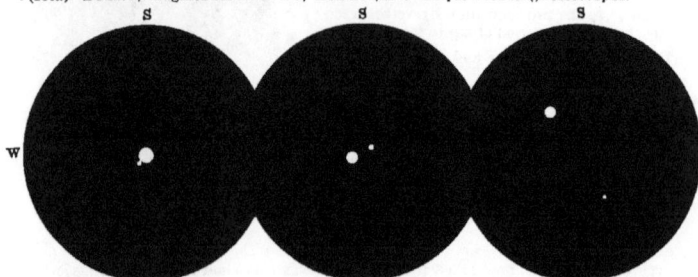

Fig. 38.—Double Star ε　　　　Fig. 39.—Double Star π　　　　Fig. 40.—Double Star ι

ξ (*Zi*)—Binary; magnitudes 4.5 and 6.5; distance, in 1880, 4″.2; yellow and red; revolution about 127 years.

κ (*Kappa*)—Double; magnitudes 5.0 and 7; distance, 12″.8; nice pair, very easy.

44 ι—Binary; magnitudes 5.0 and 6; distance, in 1880, 4″.8; revolution, 261 years.

P. XIV, 69—Double; magnitudes 5.3 and 6.8; distance, 6″.1; nice pair.

39—Double; magnitudes 5.6 and 6.5; distance, 3″.6; very easy pair.

δ (*Delta*)—Double; magnitudes 3.4 and 8.5; distance, 110″.

μ (*Mu*)—Trinary; magnitudes 4.4-7 and 8; distances, 108″ and 0″.7; the companions revolve around each other in 280 years, and around μ (*Mu*) in 120,000 years (Flam., Les Etoiles, page 142).

ζ (*Zeta*)—Double; suspected binary; magnitudes 3.6 and 4.2; distance, 0″.9; difficult pair.

OΣ 298—Binary; magnitudes 7 and 7.4; distance, 0″.2; revolution, 69 years.

CORONA BOREALIS.

Corona Borealis, or the Northern Crown, is a very characteristic constellation, and very little imagination is necessary to form the shape of a crown.

Ovid said that Ariadne, abandoned by Theseus on a deserted shore, was crying bitterly when Bacchus came to her rescue, detached her crown and threw it in the heavens; the jewels were changed into stars and formed a crown among the constellations between Hercules and the Serpent.

α (*Alpha*) is also called Gemma (the Jewel) and Alphecca.

DESIGNA-TION.	MAGNI-TUDE.	POSITION		DESIGNA-TION.	MAGNI-TUDE.	POSITION	
		R. A. 1880 h. m.	DECL. ° ′			R. A. 1880 h. m.	DECL. ° ′
α dbl.	2.2	15.30	+27. 7	ξ	5.3	16.17	+31.11
β	3.8	15.23	29.31	ο	6.0	15.15	30. 2
γ bin.	3.7	15.38	26.40	π	6.0	15.39	32.53
δ	4.2	15.44	26.27	ρ	5.8	15.57	33.41
ε dbl.	4.0	15.53	27.14	σ trin.	6.0	16.10	34.10
ζ bin. (?)	4.5	15.35	37. 2	τ	5.0	16. 5	36.47
η bin.	5.3	15.18	30.43	υ qdl.	5.8	16.12	29.28
θ	4.5	15.28	31.46	R	6.V. org.	15.44	28.31
ι	4.8	15.57	30.11	S	7.V. red	15.17	31.48
κ	4.5	15.47	36. 2	Temp. of 1866	....	15.54	26.16
λ	6.0	15.51	38. 6	U	7.V.	15.13	32. 5
μ	5.2	15.31	39.25	V	7.V.	15.45	40. 5
ν dbl.	5.0 org.	16.18	34. 2				

NOTES.

α (*Alpha*) *Alphecca*—Is a double star.

ζ (*Zeta*)—Double; magnitudes 4.5 and 6; distance, 6″.4; white and green; suspected binary.

σ (*Sigma*)—Binary; magnitudes 6.0 and 7; distance, in 1880, 3″.5; revolution, 846 years; in 1830 the companion was at 1″ only; it is a trinary in larger instruments.

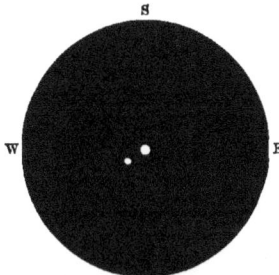

Fig. 41.—Double Star ζ

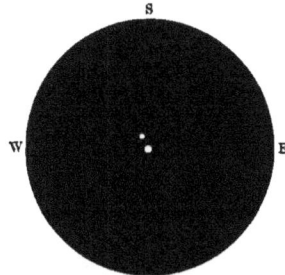

Fig. 42.—Binary Star σ

η (*Eta*)—Binary; magnitudes, 5.3 and 5.5; distance, in 1880, 0″.6; one of the most rapid orbital motion; revolution, 41 years; difficult pair.

T (1866)—There are several variables in this constellation, but the principal is T, the temporary of 1866; it is situated at 58′ S. of ε (*epsilon*). We give here a table of the variations of this remarkable star which tells its own story:

1866	MAG.	1866	MAG.	1866	MAG.	1866	MAG.
May 12	2	May 18	4.9	May 24	7.8	Aug. 1	9.7
May 13	2.5	May 19	5.3	May 26	8.0	Sept. 1	9.3
May 14	3.0	May 20	6.0	May 29	8.4	Sept. 14	8.0
May 15	3.6	May 21	6.5	June 7	9.0	Oct. 1	7.7
May 16	4.0	May 22	7.3	June 19	9.5	Oct. 15	7.5
May 17	4.5	May 23	7.5	July 1	9.7	Nov. 5	7.9

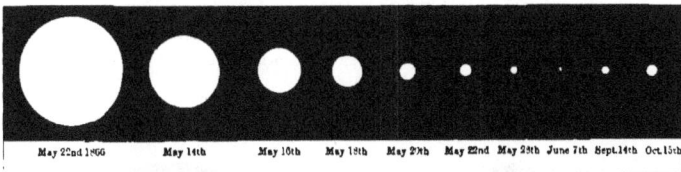

May 22nd 1866   May 14th   May 16th   May 18th   May 20th   May 22nd May 28th   June 7th   Sept.14th   Oct.15th

Fig. 43.—Diagram showing the variations of magnitude of the Temporary of 1866.

The spectral analysis has indicated all the phenomena of a veritable conflagration, which did not last more than one month; but the distance of this star is so great that the light must have taken no less than 587 years to reach us, and this *big fire* took place in 1279, more or less, according to the distance, and was seen from here only in 1866 (Flammarion, Les Etoiles, pages 143 to 149). It is now near ε (*epsilon*) in the direction of π (*pi*) of Serpens.

R.—Varies from 5.8 to 13th magnitude in 323 days; it declines faster than it increases, being sometimes below the 10th magnitude in three-fourths of the entire period; it is a little north of δ (*delta*).

V.—Varies from 7.6 to 8.8 in 3 days 10 hours and 51 minutes; it is one of the shortest period of variation; it is between η (*eta*) and δ (*delta*) of Bootes.

AURIGA.

Auriga, or the Charioteer, is one of the oldest constellations and was marked in the astronomical sphere of Eudoxus. It was named from Erichton, king of Athens, by the Greeks; and its brightest star, Capella, is the goat, Amalthea, which nursed Jupiter; the little stars, ε (epsilon), ζ (zeta) and η (eta), are called "The Kids."

DESIGNA-TION.	MAGNI-TUDE.	POSITION R.A 1880 h. m.	POSITION DECL. ° '	DESIGNA-TION.	MAGNI-TUDE.	POSITION R.A. 1880 h. m.	POSITION DECL. ° '
a (Capella)	1.3	5. 8	+45.52	φ	6.6	5.20	+34.22
β dbl.	2.3	5.51	44.56	χ	5.7	5.25	32. 6
γ (see β Tauri.)				58 ψ7	5.3	6.42	41.54
δ	4.2 yel.	5.50	54.17	46 ψ1	6.0	6.15	49.21
ε	3.V.	4.53	43.39	50 ψ2	6.0	6.31	42.34
ζ	4.0 org.	4.54	40.54	55 ψ4	5.5	6.34	44.37
η	4.0	4.58	41. 5	ψ10	5.8	6.48	45.21
θ dbl.	3.4	5.52	37.12	4 ω dbl.	5.8	4.51	37.43
ι	3.5	4.49	32.58	2	5.4	4.44	36.31
κ	5.6	6. 8	29.33	9	5.5	4.57	51.27
λ dbl.	5.5	5.11	39.59	14 trip.	5.3	5. 7	32.33
μ	6.0	5. 5	38.20	16 dbl.	5.7	5.10	33.16
ν dbl.	4.6	5.43	39. 7	41 dbl.	6.3	6. 3	48.44
ξ	5.0	5.45	55.41	63	5.9	7. 3	39.31
ο	5.9	5.37	49.47	R	7.V. red	5. 8	53.27
π	5.V. org.	5.51	45.56	*	8.0 red	4.44	28.19
ρ	6.2	5.13	41.41	*	6.8 red	4.52	39.28
σ	6.3	5.16	37.16	*	6.3 org.	6.28	38.32
τ trip.	5.5	5.41	39. 9	M. 37	cl.	5.44	32.31
υ	5.5	5.43	37.17	M. 38	cl.	5.21	35.44

NOTES.

a (*Alpha*) *Capella*—Is one of the few stars offering a parallax; found by Peters in 1842, 0″.046; which represents a distance 4,484,000 times the distance of the earth from the sun, or 420 trillions of miles; it takes the light 71 years and 8 months to come from Capella to us. Struve, in 1855, gave for the parallax 0″.305±0′.043, and Elkin, of Yale College, in 1887-88 gave one of 0″.107±0″.047; this last one, offering more precision than the above, would reduce the distance to 170 trillions of miles, and the time of the light to 29 years. This illustrates how delicate and difficult the problem of parallax is, and what a difference in the distance a little fraction of a second produces. (Revue d'Ast., 1889; page 446.)

β (*Beta*) *Menkalinan*—Mr. Pritchard, in 1891, gave two measures of the parallax of this star, 0″.065±0″.024, 0″.059±0″.025; or an average=0″.062.

14—Triple; magnitudes 5.3-7.5 and 11; distance, 15″ and 12″; the first two form an easy pair. The third one is only visible in large telescopes.

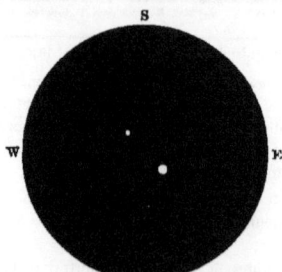

Fig. 44.—Triple Star 14.

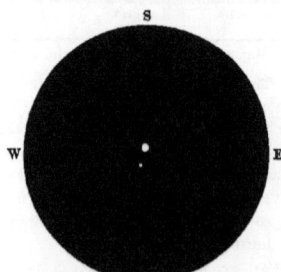

Fig. 45.—Double Star 4 ω

4 ω (*Omega*)—Double; magnitudes 5.8 and 8; distance, 6″.3; delicate pair.

M. 37—Is a cluster of over 500 stars of the 10th to the 14th magnitude; very nice and interesting with a small telescope; see about half way between β (*beta*) of Taurus and θ (*theta*).

M. 38—Is also a cluster, cross-shaped, which contains several nice small double stars; see near χ (*chi*) in the direction of Capella.

January 24th, 1892, at 2 a. m., Mr. T. D. Anderson saw a star of 5th magnitude between β (*beta*) Tauri and χ (*chi*) Aurigæ. The 1st of February he sent a postal card to Mr. Copeland, of Edinburgh Observatory, announcing its discovery. Prof. Copeland saw it also with an opera glass and telegraphed the news to the principal astronomers. Mr. Pickering, of Harvard College, examined it and said: "Copeland's Nova bright on photograph December 10th; faint, December 1st; maximum, December 20th; spectrum unique." Mr. Huggins remarks that the character of this spectrum is similar to the spectra of the Temporaries of 1866 and 1876. Position, in 1892: R. A. 5h. 25m. 4s; Decl. +30° 21′.

LYNX.

This constellation was named the Lynx, by Hevelius, in 1690, because he said that it requires a very good eyesight to see it.

DESIGNA-TION.	MAGNI-TUDE.	POSITION R. A. 1880 h. m.	DECL. ° ′	DESIGNA-TION.	MAGNI-TUDE.	POSITION R. A. 1880 h. m.	DECL. ° ′
40 α dbl.	3.4 red	9.14	+34.54	19 trip.	5.4	7.13	+55.30
38 dbl.	3.8	9.11	37.19	24	5.5	7.33	58.58
31	4.4	8.14	43.34	P. IX, 115	5.5	9.27	40. 9
21	4.7	7.18	49.27	18	5.7	7. 5	59.51
15 bin.	5.2	6.46	58.35	14 bin. (?)	5.8	6.42	59.35
2	5.5	6. 8	59. 3	Fl. 1010	6.0	7. 8	52.20
27	5.7	7.59	51.51	20 dbl.	7.5	7.13	50.22
12 trin.	5.6	6.35	59.33	R	7.V. org.	6.51	55.30
36	5.5	9. 7	43.44	*	6.2 org.	8. 0	58.36
P. VII, 169	5.5	7.31	50.43				

NOTES.

19—Triple; double with small instruments; magnitudes 5.4 and 7; distance, 14″; very easy and nice pair.

20—Double; magnitudes 7.5 and 7.5; distance, 15″; elegant pair, of equal magnitude.

38—Double; magnitudes 3.8 and 7; distance, 2″.8; very close pair.

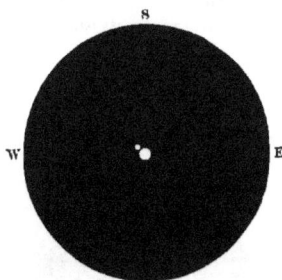

Fig. 46.—Double Star 38.   Fig. 47.—Trinary 12.

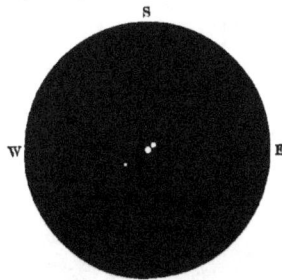

12—Trinary; magnitudes 5.8–6.5 and 7.5; distances, in 1880, 1″.4 and 8″.3. The first two revolve around each other in 676 years; the third one must take several thousand years to accomplish its revolution.

15—Is a binary, but it requires a strong power; the components are gold-yellow and azure. In 1868 Baron Dembowski, near Milan, observed the yellow sun covering the blue one about one-fourth of its diameter; an occultation of this kind is very rare; since 1872 they are going away from each other; in 1880 they were 0″.5 apart (Flam., Les Etoiles, page 165).

PEGASUS.

Pegasus is "The Winged Horse," sprung from the blood of Medusa, when Perseus killed her before going to the rescue of Andromeda; he became the horse of Jupiter and was carrying the thunder and lightning of this god; afterward he was attached to the service of the Muses, and it is he who founded the "Hippocrene Fountain," where poets went to exalt their imagination by drinking its waters.

DESIGNA-TION.	MAGNI-TUDE.	POSITION		DESIGNA-TION.	MAGNI-TUDE.	POSITION	
		R. A. 1880	DECL.			R. A. 1880	DECL.
		h. m.	° '			h. m.	° '
α dbl.	2.0	22.59	+14.34	1 dbl.	4.4	21.17	+19.17
β dbl.	2.4 red	22.58	27.26	2	4.9	21.25	23. 6
γ dbl.	2.5	0. 7	14.31	3 dbl.	6.0	21.32	6.04
ε dbl.	2.8 yel.	21.38	9.20	9	4.3	21.39	16.48
ζ dbl.	3.3	22.35	10.12	14	5.0	21.44	29.37
η dbl.	3.0	22.37	29.36	16	5.6	21.48	25.22
θ	3.6	22. 4	5.36	31	4.8	22.16	11.38
ι	4.0	22. 1	24.45	32	5.0	22.16	27.44
κ dbl.	4.0	21.39	25. 5	55	4.9 org.	23. 1	8.46
λ	4.2	22.41	22.56	56	5.0	23. 1	24.50
μ	4.3	22.44	23.58	57 dbl.	5.4 org.	23. 3	8. 2
ν	5.3	22.00	4.28	58	5.7	23. 4	9.11
ξ	4.8	22.41	11.34	59	5.4	23.06	8. 4
ο	5.0	22.36	28.41	70	5.2	23.23	12. 6
π dbl.	4.2	22. 4	32.35	77	5.5 red	23.37	9.40
ρ	5.3	22.49	8.11	78	5.2 org.	23.38	28.42
σ	5.3	22.46	9.12	85 bln.	6.0	23.56	26.27
τ	4.9	23.15	23. 5	R	7.V. red	23. 1	9.54
ν	4.9	23.19	22.43	S	7.V. org.	23.14	8.16
φ	6.0	23.46	18.27	M. 15	d.	21.24	11.38
χ	5.6	0. 8	19.33				
ψ	4.3 red	23.52	24.29				

NOTES.

The first three stars of this constellation are also called α (alpha) Markab, β (beta) Scheat and γ (gamma) Algenib, and form with Alpherat (α (alpha) Andromedæ) the "Square of Pegasus." ε (Epsilon) is named Enif and ζ (zeta) Homan.

π (Pi)—Double; magnitudes 4 and 5; distance, 12′ (minutes); it is a very easy pair of the same type as Alcor and Mizar; an opera-glass will separate it.

ε (Epsilon)—Double; magnitudes 2.8 and 9; distance, 2′ 18″; very easy in a small telescope with a large field.

1—Double; magnitudes 4.4 and 9; distance, 36″; easy pair; yellow and lilac.

3—Double; magnitudes 6 and 8; distance, 39″; very easy pair. Another delicate pair is seen in the same field.

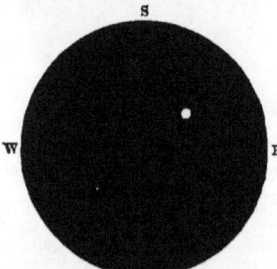

Fig. 48.—Double Star 1.

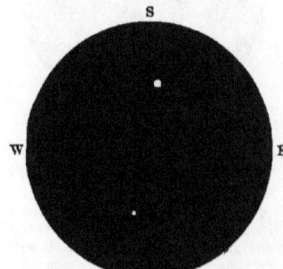

Fig. 49.—Double Star 3.

85—Binary; magnitudes 6 and 11; distance, in 1878, 0″.67; at the end of 1883 they were in contact, and the different measures taken by Mr. Burnham, at Chicago, give a period of 22 years. There is also in the same field a star of the 9th magnitude which was found at a distance of 30″ by Argelander in 1855; at 14″ by Flammarion in 1870; at 15″ at the end of 1879, and at 17″.3 at the end of 1882 by Burnham; it has kept going away ever since. Brunnow found for the parallax of the star 85 0″.054, which represents 3,805,000 times the distance of the sun from the earth, or 320 trillions of miles; the light, traveling at the rate of 190,000 miles per second, would take no less than 34½ years to reach us (Revue d'Ast., 1884; page 176).

κ (*Kappa*)—Is a binary, offering a history similar to the above. Mr. Burnham found the companion at 0″.27 on the 12th of August, 1880; at 0″.10 in July, 1890, at Lick Observatory; and for the period of revolution about 11 years, the shortest known so far. The other companion of κ (*kappa*), of the 9th magnitude, is at 12″ distance in 1891 (Revue d'Ast., June, 1891; page 210).

R. and S.—Vary from the 7th to the 12th magnitude, but require powerful telescopes.

M. 15—Is a cluster of several hundreds of stars, visible with a small power; marked on our planisphere N. W. of ε (*epsilon*).

Fig. 50.—Cluster Messier 15.

EQUUELUS.

Equuelus, or "The Little Horse," is a constellation formed by Hipparchus about 130 years B. C., and contains only a few small stars.

DESIGNA-TION.	MAGNI-TUDE.	POSITION		DESIGNA-TION.	MAGNI-TUDE.	POSITION	
		R. A. 1880 h. m.	DECL. ° ′			R. A. 1880 h. m.	DECL. ° ′
α	4.0	21.10	+4.44	δ bin.	4.5	21. 9	+9.31
β qdl.	5.0	21.17	6.18	1 ε trin.	5.4	20.53	3.50
γ dbl.	4.5	21. 5	9.38	2 ζ dbl.	6.3	20.56	6.42

NOTES.

γ (*Gamma*)—Double; magnitudes 4.5 and 6; distance, 6′ 6″; very easy with an opera glass; the comet of 1680 passed close by on the 3d of January, 1681; Knott, in 1867, discovered a companion at 2″.

ε (*Epsilon*)—Trinary; magnitudes 5.4-7.5 and 7.5; distances, 11″ and 0″.9. In 1835 Struve discovered for the first time the two companions at 0″.35 only from each other; it is a very important system.

δ (*Delta*)—Binary; magnitudes 4.5 and 5; distance, 0″.2 in 1880; very rapid orbital motion, time about 12 years. There is another companion of the 10th magnitude, measured at a distance of 20″ in 1781; at 27″ in 1835; at 30″ in 1847; at 33″ in 1859; at 34″ in 1870; at 38″ in 1880. The change in the distance of this star is due to proper motion.

β (*Beta*)—Is also a multiple star.

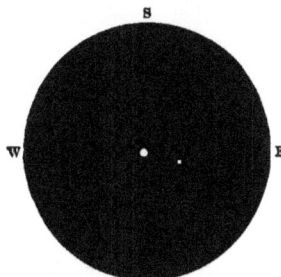

Fig. 51.—Trinary 1 ε

DELPHINUS.

Delphinus, the Dolphin, appeared in the sphere of Eudoxus. Who is he? Some say that he is the Dolphin who saved the poet Arion from shipwreck; others that he is the one that Neptune sent to the discovery of Amphitrite; others that he is the pirate Acetes, who protected Bacchus; others that it represents Apollo coming back from Crete, and some others claim that it is the fish in which Jonah remained for three days and three nights.

DESIGNA-TION.	MAGNI-TUDE.	POSITION			DESIGNA-TION.	MAGNI-TUDE.	POSITION		
		R. A. h. m.	1880	DECL. ° ′			R. A. h. m.	1880	DECL. ° ′
α dbl.	3.7	20.34		+15.29	ι	5.7	20.32		+10.57
β trip.	3.3	20.32		14.11	κ dbl.	4.8	20.33		9.40
γ bin.	3.4	20.41		15.42	Σ 2703 dbl.	7.5	20.31		14.23
δ	4.0	20.38		14.38	R	8.V. org.	20. 9		8.44
ε	4.0	20.27		10.54	S	8.V. org.	20.38		16.39
ζ	4.9	20.30		14.16	T	8.V. org.	20.40		15.58
η	5.8	20.28		12.37	*	6.8 org.	20.40		17.39
θ	6.0	20.33		12.54	*	7.0 org.	20.32		17.51

NOTES.

α (*Alpha*) is called also Sualocin and β (*beta*) Rotanev. The letters of these names, reversed, form Nicolaus Venator, the Latin name of Niccolo Cacciatore, who was connected with the Palermo Observatory.

γ (*Gamma*)—Double; binary; magnitudes 3.4 and 6; distance 11″; orange and green, very nice; the companion changes color; sometimes orange, sometimes yellow, sometimes blue, sometimes green; but generally emerald green. Period, 26 years.

κ (*Kappa*)—Double; magnitudes 4.8 and 11; distance, 10″; easy pair, but the companion is small.

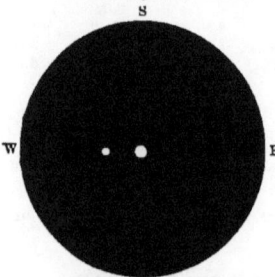

Fig. 52.—Double Star γ

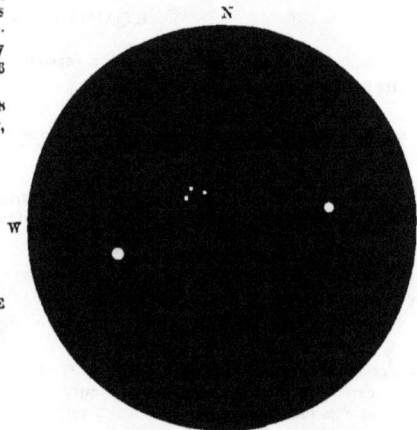

Fig. 53.—Triple Star Σ 2703, between β and ζ

β (*Beta*)—Triple, even quadruple, having two companions of 10th and 13th magnitude at 35″ and 28″, and the star β (*beta*) being a very close binary, distance 0″.4 only; in rapid orbital motion, discovered by Mr. Burnham in 1873, its period about 26 years.

Σ 2703—Triple; magnitudes 7.6–7.6 and 7.8; distance, 26″ and 69″; beautiful triple between β (*beta*) and ζ (*zeta*). These three stars, β (*beta*), Σ 2703 and ζ (*zeta*), appear in the field of a small telescope and form a nice group (Fig. 53). Image not reversed.

θ (*Theta*)—If you also look at θ (*theta*) you will find a nice field of small stars.

OΣ 527—Is also a binary; magnitudes 7 and 8; distance, 0″.5; period of revolution, 54 years; too close for common telescopes.

There are several variables in this constellation, but they require good telescopes and large power to be followed.

CYGNUS.

Cygnus, according to the Greeks, is Jupiter himself, taking the form of a swan to seduce the innocent Leda. Ovid said that he was a relative of Phaeton changed into a swan by Apollo, after the fall of Phaeton from the heavens; Hipparchus and Ptolemy called it simply Ornithos (the Bird); Manetho, the Hen; the Arabs, the Pigeon; but lately the name Swan is generally recognized: nevertheless it is a very old constellation, and was already in Eudoxus' sphere.

DESIGNA-TION.	MAGNI-TUDE.	POSITION		DESIGNA-TION.	MAGNI-TUDE.	POSITION	
		R. A. 1880 h. m.	DECL. ° '			R. A. 1880 h. m.	DECL. ° '
a dbl.	2.0	20.37	+44.51	19	5.6	19.45	+38.25
β dbl.	3.4 yel.	19.26	27.43	20 d	5.5	19.48	52.42
γ	2.5	20.18	39.52	27 b1	5.3	20. 2	35.39
δ bin.	2.9	19.41	44.50	28 b2	5.0	20. 5	36.29
ε	2.7 yel.	20.41	33.31	29 b3	5.6	20.10	36.26
ζ dbl.	3.3	21. 8	29.44	32	5.5 red	20.12	47.21
η	4.6	19.52	34.46	33	4.4	20.11	56.12
θ	4.6	19.33	49.56	34 P (1600)	5.5	20.13	37.40
ι	4.0	19.27	51.28	39	5.0 red	20.19	31.48
κ	4.1	19.14	53. 9	41	4.8	20.25	29.57
λ bin.	5.3	20.43	36. 3	47	5.2 org.	20.30	34.51
μ trip.	4.6	21.39	28.12	48	5.5	20.33	31. 9
ν	4.2	20.53	40.42	52 dbl.	4.6	20.41	30.16
ξ	4.1 red	21. 1	43.26	59 trip.	5.0	20.56	47. 3
30 qdl.	4.0 red	20.10	46.24	61 bin.	5.4	21. 2	38.10
32	4.5	20.12	47.21	68 A	5.0	21.14	43.27
π1	4.8	21.38	50.38	70	5.5	21.22	36.35
π2	4.5	21.42	48.45	71 S	5.4 red	21.26	46. 0
ρ	4.2	21.30	45. 3	72	5.5	21.30	38. 0
σ	4.4	21.13	38.54	74	5.5	21.32	39.53
τ	4.0	21.10	37.32	16 c dbl.	6.0	19.39	50.15
υ qdl.	4.6	21.13	34.23	Σ 2486 dbl.	6.6	19.00	49.37
φ	5.0	19.35	29.52	Temp. of 1876	....	21.37	42.18
χ1 dbl.	5.3	19.42	33.28	R	7.V. org.	19.34	49.56
χ2	5.V. red	19.46	32.37	T	5.V.	20.42	33.56
ψ dbl.	5.3	19.53	52. 7	U	7.V. red	20.16	47.31
ω1	6.0 yel.	20.23	48.59	*	8.0 red	19.36	32.21
ω2 dbl.	5.0	20.26	48.33	*	6.7 red	21.32	44.50
ω3 dbl.	5.9 yel.	20.28	48.49	*	6.0 org.	20.13	40. 0
				*	6.3 org.	20.18	40.39
2	5.3	19.20	29.24	*	6.3 org.	20.49	32.59
3	6.0 red	19.20	24.42	*	6.5 org.	20. 2	34.34
4	5.0	19.22	36. 6	*	6.7 org.	19.57	36.46
8	5.0	19.27	34.13				

NOTES.

a (Alpha) Deneb—For this star a negative parallax (−0″.042±0″.047) was obtained by Mr. Elkin, of Yale College, in 1887, and its distance can not be calculated.

χ1 (Chi)—Double; magnitudes 5.3 and 8; distance 26″; near it, 15′ S. S. W., there are two stars of the 8th magnitude at 3″ only, revolving around each other and going in the same direction; they most likely form a part of the same system. (See χ (chi).)

χ2 (Chi)—The star below χ (chi) on our planisphere is a remarkable variable, varying from the 4.5 to the 13th magnitude in a period of 406 days, with some irregularities; it was observed in 1687 by G. Kirch, and the period fixed by Maraldi; it is a sun which sends 4,600 times more light and heat at the first magnitude (4.5) than it does at the second (13th) magnitude.

Fig. 54.—Diagram showing the periodical variations of χ.

34 P.—Between γ (*gamma*) and η (*eta*); was first seen by Blaeu, August 18th, 1600, and noted 3d magnitude; in 1622 it was of the 5th magnitude; from 1655 to 1660 Cassini saw it of 3d magnitude; and the 31st October, 1660, it came down to the 5.5; from 1662 to 1666 it was not visible to the naked eye; it was noted in 1667-82 and 1715 of the 6th magnitude; in 1793 and 1807 Piazzi saw it of the 5th magnitude; Pigott gave it a period of variability of 18 years, but to-day it seems to be stationary at 5.5.

T. (1670)—West of Albireo, Father Anthelme, at Dijon, saw a star of 3d magnitude the 20th of June; the 11th of July it was of the 4th magnitude; August 10th of the 5th, and it kept going down; but on March 17th, 1671, it was found of the 4th magnitude; in April and May Cassini noted it brighter than β (*beta*); then it came down again so quickly that at the end of August it was no more visible to the naked eye; in 1672 Hevelius saw it of 3d magnitude, but it disappeared again in September of the same year; since then nobody has seen it. Flammarion adds: There is at less than one minute of the position given for this star a variable of from 8.6 to 9.3 magnitude (S in Vulpecula); perhaps it is the temporary of 1670? watch it! (See between Albireo and 15 of Vulpecula.)

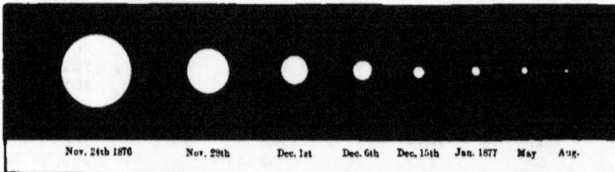

Fig. 55.—Diagram showing the variations of the Temporary of 1876.

T (1876)—Near ρ (*rho*), about on the line passing through α (*alpha*) and ξ (*zi*), Mr. Schmidt, of Athens, on the 24th of November, 1876, saw a star of the 3d magnitude; he was observing this region four days previously, but did not notice anything there; it soon came down, and on the 5th of December it was of the 5th magnitude; Father Secchi found it of 7th magnitude the 5th of January, 1877; its spectrum confirmed the idea of a fierce conflagration; in September, 1877, Lord Lindsay found that it had all the appearance of a nebula; it is now of the 12th magnitude. It is a remarkable fact that mostly all temporaries are in the region of the "Milky Way."

There are a great many more variables in this constellation; we will note only:

R.—In the same field with θ (*theta*), which varies from the 7th to 14th magnitude in 405 days.

T.—Near ε (*epsilon*); varies from 5th to 6th irregularly; visible with an opera glass.

S.—Varies from 9th to 13th in 322 days.

U.—Varies from 7 to 10.5 magnitude in 465 days.

β (*Beta*) Albireo—Double; magnitudes 3.5 and 6.0; distance, 34"; gold-yellow and sapphire; easy pair, and one of the finest.

o² (*Omicron*)—Is quadruple; triple in small telescope; magnitudes 4.3-7.5 and 5.5; distances, 107" and 338"; the principal is yellow, the companions blue; very easy with an opera glass; nice field of small stars.

ψ (*Psi*)—Double; magnitudes 5.3 and 8; distance, 3".5.

μ (*Mu*)—Triple; magnitudes 4.6-6.0 and 7.5; distances, 3".7 and 210"; the third one forms only a group of perspective.

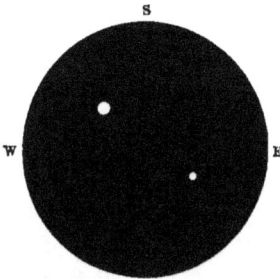

Fig. 56.—Double Star β

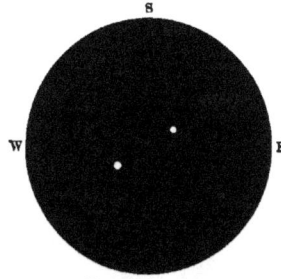

Fig. 57.—Binary 61.

61—Binary; magnitudes 5.5 and 6.0; distance, 20″; rapid motion, 5″.16 per year toward σ (*sigma*), near which it will be in 1,500 years; 4,000 years ago it was near ε (*epsilon*); if all the stars had such motion, all the constellations would change in less than 1,000 years. The parallax of this star has been determined many times from 1838 by Bessel to 1888 by Belopolsky, and fixed as the result of all determinations at 0″.44 (Mr. Asaph Hall, of Washington Observatory, gave 0″.270±0″.010—far away from all other observers). It may be interesting to make a table of all the measures:

AUTHORS.	PARALLAX.	AUTHORS.	PARALLAX.
Bessel, 1838–40	0″.348±0″.010	O. Struve, 1852–53	0″.506±0″.028
Auwers, 1868	0″.423±0″.013	J. Lamp, 1883	0″.506±0″.023
C. A. F. Peters, 1842–43	0″.349±0″.080	Auwers, 1860–62	0″.564±0″.016
Johnson, 1852–53	0″.402±0″.016	R. Ball, 1877–78	0″.465±0″.050
Socolof, 1863–66	0″.437±0″.069	R. Ball, 1877–78	0″.468±9″.032
	0″.376±0″.044	Asaph Hall, 1880–86	0″.270±0″.010
	0″.486±0″.080	Pritchard, 1886–87	0″.433±0″.014
	0″.493±0″.055	Belopolsky, 1888	0″.535±0″.092

The parallax adopted, 0″.44, represents 469,000 times the distance of the earth from the sun, or 40 trillions of miles; the light takes 7 years and 3 months to reach us; it is the nearest star visible to the naked eye from our country; α (*alpha*) Centauri, being too far south to be seen from our latitudes, is the only one nearer to our system. (Revue d'Ast., 1889.)

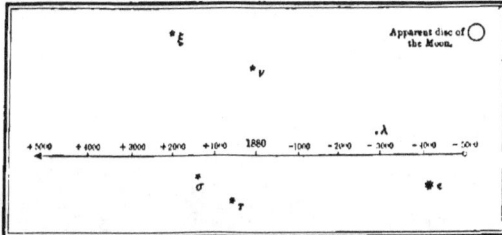

Fig. 58.—Diagram showing the motion of the star 61 for 10,000 years.

16 c.—Double; magnitudes 6.0 and 6.5; distance, 37″; easy pair.

δ (*Delta*)—Binary; magnitudes 2.9 and 8; distance, 1″.6; white and blue; the companion is a variable, but it is a difficult pair; time of revolution, 336 years.

52—Double; magnitudes 4.6 and 9; distance, 7″; the first is a variable orange, 4th to 6th magnitude, and the companion is blue.

Σ 2486—Binary; magnitudes 6 and 6.5; distance, in 1885, 10″; it is also in rapid motion. Mr. Ball, from observations taken in 1880–81, gives for its parallax 0″.482±0″.054, which represents a distance of 39 trillions of miles, or 428,000 times the distance of the sun from the earth; the light takes over 6 years and 8 months to reach us (Revue d'Ast., 1885; page 312). Prof. Hall, in 1885, found a negative parallax (−0″.021), and the distance is doubtful, so far. (See S. W. of θ (*theta*).)

VIA LACTEA.

Via Lactea, the Galaxy, or "Milky Way"—Is composed entirely of suns like our own, millions and millions of miles distant from each other.

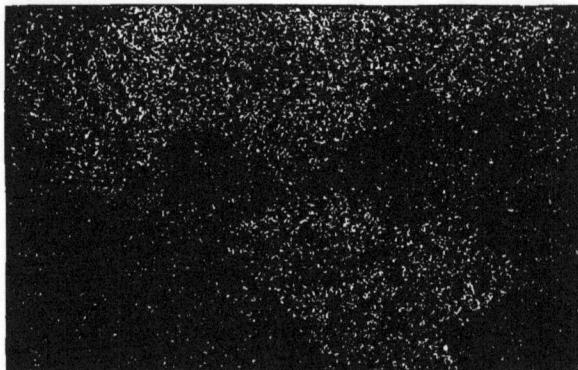

Fig. 59.—A portion of the Milky Way in Scutum Sobiesii.

In contemplating the heavens at night, especially in summer, when the sky is very clear and the moon is invisible, we can see it extending from one point of the horizon to the other, describing nearly half of a circumference above us, and if we could see through the earth it would appear at the antipodes, making a complete circle; consequently the earth, the sun and the other planets are in its center or some other part of it, but undoubtedly in it. Its light is not uniform; when the stars are close together and numerous the region is very bright, and the brightness depends upon their number; some parts are poor and some are entirely without stars.

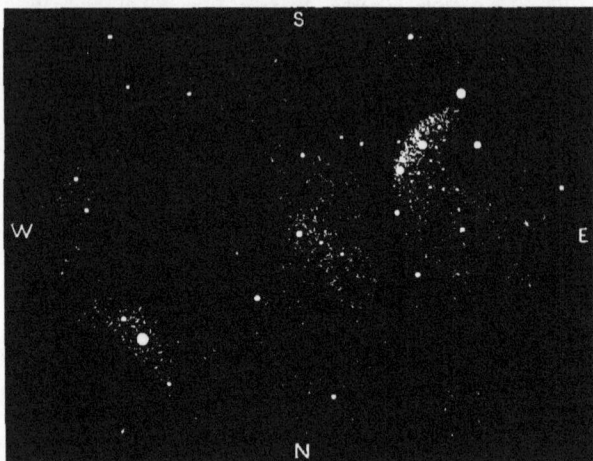

Fig. 60.—Holes in the Milky Way.   From a sketch by Mr. Trouvelot.

Fig. 60, taken from a sketch made by Mr. Trouvelot, is a miniature of the Coal Sacks, and could be seen in constellation Sagittarius, between γ (gamma) and the star 3 X.

There is a very white region north of Aquila; another in Scutum Sobiesii; another in Sagittarius; three more near α (alpha), β (beta) and γ (gamma) in Cygnus; another in Perseus. There is a dark region between α (alpha) and β (beta) in Cassiopeia; another in Cygnus, etc. Its width is not regular; in Cygnus it separates into two branches, the principal one passes through Aquila, Scutum Sobiesii, Sagittarius and Scorpius; the other goes through Ophiuchus, the Serpent, and seems to fade away in Scorpius; they reunite in Centaurus. In Triangulum Australis it is very bright, passes over the Southern Cross, where is seen a black hole in pear shape, 8° by 5°, called the "Coal Sack," containing only a small star of the 6th magnitude, hardly visible to the naked eye; then it comes to its narrowest size, measuring only 4° in width, when it has no less than 16° in Cygnus and 22° between Ophiuchus and Antinous; it gets wider in Argo Navis; touches Canis Major, Monoceros, Orion, Gemini and Taurus; becomes irregular in Auriga; passes through Perseus, Andromeda and Cassiopea, and then comes back to the starting point in Cygnus.

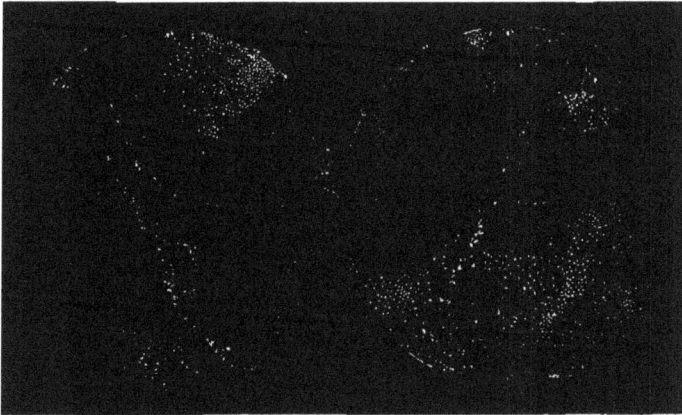

Fig. 61.—The Milky Way and the Nebulæ (from R. A. Proctor).

Notwithstanding the beautiful researches of W. Herschel, John Herschel, W. Struve, Maedler, Secchi and Richard Proctor, it would be premature to give the shape of the Galaxy, but as its density seems to be about twice as great near the 18th hour of right ascension as at the 6th hour, we have very good reasons to believe that we are nearer to Sirius than to Scutum Sobiesii (Flam., Les Etoiles, page 184). It was in trying to figure out the number of stars of the "Milky Way" that W. Herschel introduced his famous method of "gauging the heavens," by the means of which he came to the conclusion that it is composed of no less than 18,000,000 of stars. Looking through his telescope and counting the stars passing through the field he figured out 116,000 of them in a quarter of an hour in the richest part of it. In Cygnus, in a field of the apparent diameter of the moon, there were from 1,800 to 2,000 stars; in Aquila some regions have as many as 2,300 for the same field, and in Scutum Sobiesii, in a space of five square degrees, the amount reaches the extraordinary number of 331,000 stars. Some regions contain only 500, 200, and 80 stars, and even, in some places in a field of 15 minutes, not a star appears. The "gauges" of W. Herschel and W. Struve give for the proportion of the stars in all the heavens:

In the "Milky Way"			122 stars for a field of 15 minutes in diameter.					
15° from it, one way or the other			30	"	"	"	"	
30°	"	"	"	18	"	"	"	"
45°	"	"	"	10	"	"	"	"
60°	"	"	"	6	"	"	"	"
75°	"	"	"	4	"	"	"	"

This plainly shows that there are over thirty times as many stars in the plan of the Galaxy as at 90° each way from it, and that their density progressively decreases according to their distance from it. The distribution of the nebulæ is just the opposite. R. Proctor made a diagram illustrating the remarkable fact that their agglomeration is the greatest at 90° each way from the " Milky Way;" outside of the Galaxy he shows 4,053 white spots, each one representing a nebula. Still, this immense cluster, of which our sun is a small star, must not be considered the largest of the universe; it would be the same illusion as if we were supposing the sun making its revolution around the earth. "If the light coming from one point at the edge of the ' Milky Way' and going in the diametrically opposite direction would take 3,000 years to reach the other side, this cluster, seen at 334 times its size, would be only 10 minutes in diameter, and the light would have to travel no less than 1,000,000 years to finish its journey" (from F. Hoefer's biography of W. Herschel). That distance is nothing for the Infinite—when we reach it we have not made a step. We certainly would not see the constellations as they appear from here, but other marvels would cause admiration. The earth is indeed very small.

VULPECULA.

Vulpecula, or the Fox, was formed in constellation by Hevellus, about 1660 A. D. He represented this little animal carrying a goose that he had stolen, and placed it between the Eagle and the Vulture (Lyra), because he says that the fox is astute, voracious and ferocious, like those birds.

DESIGNA-TION.	MAGNI-TUDE.	POSITION		DESIGNA-TION.	MAGNI-TUDE.	POSITION	
		R. A. 1880 h. m.	DECL. ° '			R. A. 1880 h. m.	DECL. ° '
6 α dbl.	4.4	19.24	24.25	28	5.4	20.33	+23.42
1 dbl.	5.0	19.11	+21.11	29	5.3	20.33	20.46
4	5.2	19.20	19.34	30	5.8	20.40	24.50
9	5.5	19.29	19.30	31	5.5	20.47	26.38
12	5.8	19.46	22.18	32	5.7	20.49	27.36
13	5.0	19.48	23.46	R	8.V. org.	20.50	23.21
15	5.5	19.56	27.26	S	8.V. org.	19.43	26.59
16 dbl.	5.7	19.57	24.37	T	6.V.	19.47	24.41
17	5.5	20. 2	23.17	H. VIII, 20	cl.	20. 7	26.21
19	6.0 org.	20. 7	26.27	M. 27	neb.	19.54	22.24
16 Hev.	5.2	20. 7	26. 8	Σ 2245 dbl.	6.5	17.51	18.21
23	5.0 org.	20.11	27.27	Temp. of 1670	....	19.43	27. 2

NOTES.

M. 27—Is the remarkable nebula called " dumb-bell," from its shape; it contains many small stars; it looks like a double nebula with a small telescope; larger instruments unite the two like a dumb-bell; still stronger power will modify it more (Figs. 62 and 63).

Fig. 63.—Nebula, M. 27, in Lord Rosse's telescope.

II. VIII, 20—Cluster, near 16; it is a cluster of 104 stars from the 9th to the 13th magnitude; nice field.

R.—Varies from 8th to 13th in 137½ days.

S.—Varies from 8.6 to 9.3 in 68 days (between 15 and Albireo). (See the temporary of 1670 in Cygnus.)

T.—Varies from 5th to 6th magnitude, and perhaps less.

SAGITTA.

Sagitta, or the Arrow, is the smallest constellation, but nevertheless one of the oldest, being already marked in Eudoxus' sphere.

DESIGNA-TION.	MAGNI-TUDE.	POSITION		DESIGNA-TION.	MAGNI-TUDE.	POSITION	
		R. A. 1880	DECL.			R. A. 1880	DECL.
		h. m.	° ′			h. m.	° ′
a	4.6	19.35	+17.44	10	6.0	19.51	16.19
β	4.5	19.36	17.11	11	6.0	19.52	16.27
γ	3.8	19.53	19.10	13	6.0	19.55	17.11
δ	4.3	19.42	18.14	15	6.0	19.59	16.45
ε dbl.	5.7	19.32	16.11	R	8.V. org.	20. 9	16.22
ζ dbl.	5.5	19.44	18.50	*	6.5 org.	20. 3	16.20
η	5.5	20. 0	19.39	II. VI, 11	cl.	19.25	20. 1
θ trip.	6.2	20. 5	20.33	M. 71	cl.	19.49	18.28

NOTES.

ε (Epsilon)—Double; magnitudes 5.7 and 8; distance, 92″; very easy pair.

ζ (Zeta)—Double; magnitudes 5.5 and 9; distance, 8″.6; these two stars are sometimes white and blue, sometimes yellow and blue, sometimes yellow and violet, sometimes yellow and red and sometimes blue and violet—very curious variations; it is also in rapid proper motion, 60″ every 100 years.

θ (Theta)—Triple; magnitudes 6.2-8 and 7; distances, 11″ and 76″; very easy for small instruments. The first two stars, generally called A and B, have the same proper motion; the third one, C, is stationary and independent of the others.

H. VI, 11—Is a nice cluster, visible with an opera glass, formed of stars from the 6th to the 10th magnitudes; a large field and small power offer a beautiful sight (see in the direction of β (*beta*) to α (*alpha*) near 9 of Vulpecula).

S

13—Is a bright star of the 6th magnitude, s h i n i n g like a golden pearl; in the field will be found a nice little star, very red, and a pretty little pair; beautiful sight.

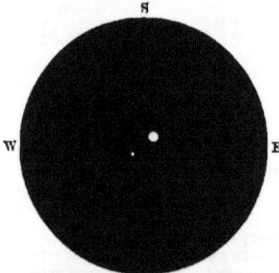

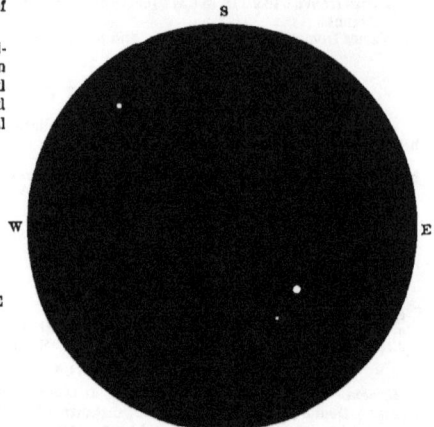

Fig. 64.—Double Star ζ   Fig. 65.—Triple Star θ

M. 71—Is also a cluster, marked in our planisphere near ζ (*zeta*).

R.—Varies from 8.6 to 10.1 magnitude in 70 days and 10 hours. (See north of ρ (*rho*) of Aquila in the direction of θ (*theta*).)

This little constellation is very rich in curiosities.

LYRA.

This constellation represents the lyre or harp of Orpheus which was given to him by Apollo; it was afterward called the Turtle, and still later, the Falling Vulture. Vega comes from the Arab Wa-ki (Al-nasr-al-waki).

DESIGNA- TION.	MAGNI- TUDE.	POSITION R. A. 1880 h. m.	DECL. ° '	DESIGNA- TION.	MAGNI- TUDE.	POSITION R. A. 1886 h. m.	DECL. ° '
α (Vega) dbl.	1.2	18.33	+38.40	μ	5.5	18.20	+39.27
β qdl.	3.V.	18.46	33.13	ν mult.	6.0	18.45	32.32
γ dbl.	3.3	18.54	32.32	16 dbl.	5.5	18.59	46.46
δ dbl.	4.4 red	18.50	36.45	13 R	4.V. yel.	18.52	43.47
ε bin.	4.4	18.40	39.31	34931	5.0	18.42	26.31
ζ dbl.	4.4	18.41	37.29	33739	5.4	18.12	42. 7
η dbl.	4.6 red	19.10	38.56	M. 56	cl.	19.12	29.59
θ dbl.	4.2	19.12	37.55	M. 57	neb.	18.49	32.53
ι	5.0	19. 3	35.55	*	6.5 red	18.39	39.11
κ	4.7	18.16	36. 1	*	8.0 red	18.28	36.54
λ	5.7 org.	18.55	31.58				

NOTES.

α (*Alpha*) *Vega*—Double; magnitudes 1.2 and 9; distance, 47"; difficult, on account of the brilliancy of Vega; it is not a binary; the little star is independent of it, and has been used to determine the parallax of Vega, which is noted 0".15 (Revue d'Ast., Dec., 1889; page 446). It represents 1,375,000 times the distance of the earth from the sun, or 12 trillions of miles; it takes 21 years and 8 months for the light to reach us. Vega is coming toward us with an approximate velocity of 44 miles per second; but as our system travels itself in that direction (toward Hercules) a part of it belongs to the sun. By the effects of precession of the equinoxes, Vega was the "Polar Star" of the earth 14,000 years ago, and will be "Polar Star" again in 12,000 years; it is one of the brightest stars.

β (*Beta*) is also called Sheliak and γ (*gamma*) Salaphat.

ε (*Epsilon*)—Is a quadruple composed of two binaries; ε¹ magnitudes 6 and 7; distance, 3″.2; revolution about 1,800 years; and ε², binary; magnitudes 5.5 and 6; distance, 2″.4; revolution about 3,700 years. The distance between ε1 and ε2 is 3′ 27″. The revolution of the two systems around their center of gravity must take about 1,000,000 years! (Flam., Les Etoiles, page 216.) Large telescopes show several small stars between the binaries.

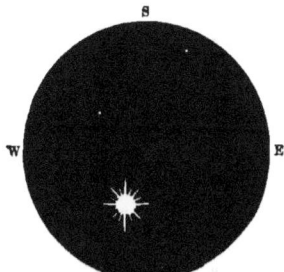

Fig. 66.—Double Star Vega.

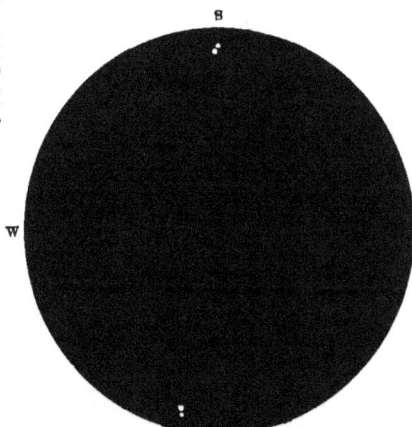

Fig. 67.—Quadruple Star ε.   Scale, 100″ — 1 inch.

ζ (*Zeta*)—Double; magnitudes 4.5 and 5.5; distance, 44″; yellow and green; one of the nicest pair; very easy.

η (*Eta*)—Double; magnitudes 4.6 and 9; distance, 28″; delicate pair; pale yellow and violet.

δ (*Delta*)—Double; magnitudes 4.5 and 5.5; distance, 12′; visible with the naked eye and opera glass; rich field.

β (*Beta*)—Varies from 3.4 to 4.5 in 12 days 21 hours and 51 minutes; Father Secchi saw its spectrum one day of its maximum similar to γ (*gamma*) of Cassiopea, indicating a fierce fire, but never noticed it after; it has three little companions in the field.

Σ 3130—Binary; magnitudes 7.4 and 11; distance, 2.″7. Principal star thought to be a close pair by Otto Struve.

13 R.—Varies from the 4th to the 5th magnitude in 46 days; Flamsteed noted it of 6th magnitude at the end of the 17th century.

M. 57—Is the famous ring nebula; it is an ellipse 78″ long, 60″ wide; Herschel called it "perforated nebula;" but in large telescopes the center is filled up with fine lines and nebulosities; it is most likely a system in formation (Flam., Les Etoiles, page 217); it is between β (*beta*) and γ (*gamma*).

Fig. 68.—Nebula, M. 57.
In common telescope.

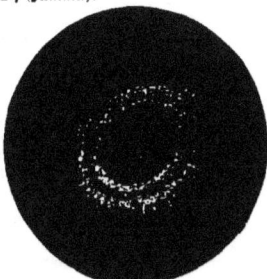

Fig. 69.—Nebula M. 57, in Lord Rosse's telescope.

Fig. 70.—Cluster, M. 56.

M. 56—Is a globular cluster, the opposite of the above; the center being very bright, and the light gradually fading away on the edges; it is formed of several hundred stars, and will be found about half way between Albireo and γ (*gamma*); diameter 3 minutes.

HERCULES.

Hercules, or "The Kneeler," is an old constellation, appearing on the sphere of Eudoxus, but the name Hercules is comparatively new. C. Flammarion noticed it for the first time in an edition of Hyginus, dated A. D. 1485.

DESIGNA- TION.	MAGNI- TUDE.	R. A. 1880 h. m.	DECL. ° '	DESIGNA- TION.	MAGNI- TUDE.	R. A. 1880 h. m.	DECL. ° '
α dbl.	3.V. org.	17. 9	+14.32	13 p	7.5	16. 9	+11.48
β dbl.	2.4 yel.	16.25	21.45	8 q	6.0	16. 6	16.59
γ dbl.	3.6	16.17	19.27	5 r	5.8	15.56	18. 9
δ bin.	3.6	17.10	24.59	s	6.0	16.46	30. 1
ε	3.5	16.50	31. 6	107 t	5.5	18.16	28.49
ζ bin.	2.9	16.37	31.49	68 u, dbl.	4.V. red	17.13	33.14
η dbl.	3.5	16.39	39. 9	72 w	5.3	17.16	32.37
θ	3.8	17.52	37.16	77 x	6.0	17.24	48.22
ι dbl.	3.7	17.36	46. 4	82 y	5.8	17.34	43.38
κ dbl.	5.5	16. 3	17.22	88 z	7.0	17.47	48.26
λ dbl.	5.0 yel.	17.26	26.12	42 trip.	4.9	16.35	49.10
μ dbl.	3.8	17.42	27.48	52	5.2	16.46	46.12
ν	4.4	17.54	30.12	53	5.8	16.48	31.54
ξ	4.0	17.53	29.16	60 dbl.	5.0	17. 0	12.55
ο	4.0	18. 3	28.45	70 dbl.	5.0	17.16	24.35
π	3.4 red	17.11	36.57	93	5.0	17.55	16.46
ρ bin.	4.0	17.20	37.16	95 dbl.	4.8	17.57	21.36
σ	4.3	16.30	42.41	96	5.0	17.57	20.51
τ	3.5	16.17	46.36	100 dbl.	6.0	18. 3	26. 5
υ	4.5	16. 0	46.23	101	5.2	18. 4	20. 2
φ	4.0	16. 5	45.16	102	4.4	18. 4	20.49
χ	4.7	15.48	42.47	109	4.2 yel.	18.19	21.43
ω	5.0	16.20	14.19	110	4.2	18.41	20.26
				111	4.0	18.41	18. 3
104 A	5.0 org.	18. 7	31.22	113	4.5	18.50	17.58
99 b	5.0	18. 2	30.34	P. XVI, 279	5.8	16.58	14.16
61 c	5.7	17. 4	36. 6	31312	5.0	17. 7	40.55
59 d	5.2	16.57	33.44	31694	5.8	17.19	40. 6
69 e	4.8	17.14	37.25	R	8.V. red	16. 1	18.42
90 f	5.2	17.49	40. 1	S	6.V. org.	16.46	15. 9
30 g	5.V. red	16.25	42. 9	T	7.V. org.	18. 5	31. 0
29 h	5.3	16.27	11.46	U	7.V. red	16.20	19.10
43 i, dbl.	5.8 red	16.40	8.48	W	8.V.	16.31	37.35
47 k	5.8	16.44	7.28	*	7.0 red	16.43	42.27
45 l	5.8	16.42	5.28	*	6.6 org.	15.59	47.34
36 m	6.0	16.35	4.28	*	7.0 org.	18.14	23.14
28 n	5.9	16.27	5.47	M. 13	cl.	16.37	36.41
21 o	6.2	16.18	7.14	M. 92	cl.	17.13	43.15

NOTES.

α (*Alpha*) *Ras Algethi*—Double variable; magnitudes 3.1 to 3.9 and 5.5; distance, 4".7; beautiful pair; orange and emerald; very easy. Mr. Jacob, in 1856-58, in using the companion, obtained a parallax=0".062±0".007; but as they form a physical pair the parallax is not sure.

β (*Beta*) *Korneforos*—Is also double.

ρ (*Rho*)—Binary; magnitudes 4.0 and 5.5; distance, in 1880, 3".7; nice pair.

κ (*Kappa*)—Double; magnitudes 5.5 and 6.4; distance, 30"; this easy pair looks like Mizar and Alcor, having in the field another star of the 6th magnitude, north of it.

95—Double; magnitudes 5.5 and 5.8; distance, 6"; gold-yellow and pale azure; very beautiful pair.

δ (*Delta*)—Double; magnitudes 3.6 and 8; distance, in 1880, 18'; change from proper motion; the first is bright blue, the companion violet.

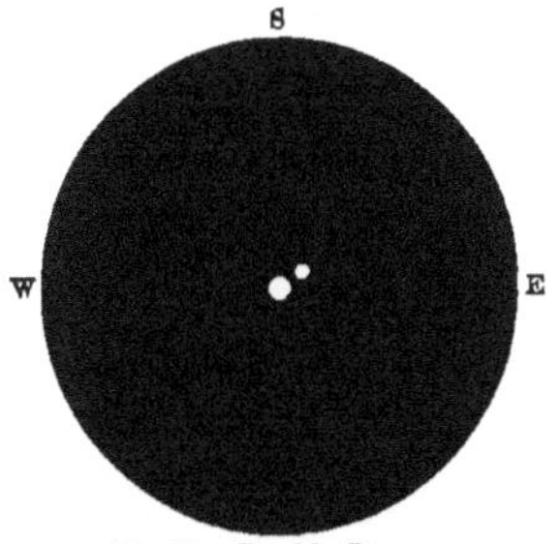

Fig. 71.—Double Star α

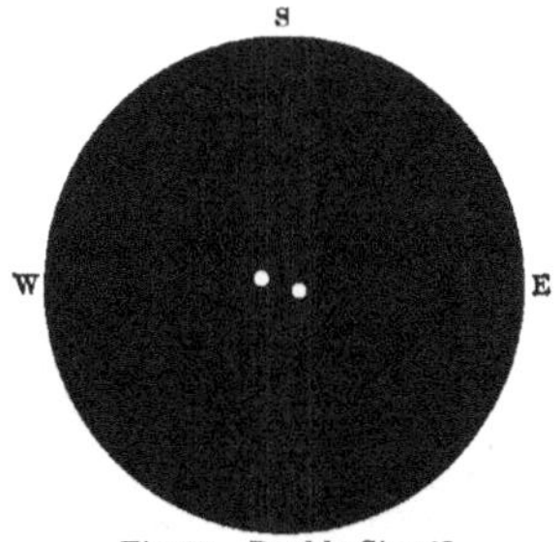

Fig. 72.—Double Star 95.

ζ (Zeta)—Binary; magnitudes 3 and 6; distance, in 1880, 1″.3; rapid orbital motion; time 34½ years; the distance varies from less than 0″.6 to 1″.5; and good power is required to separate it. The companion disappears for about three years at each revolution, when it comes closer than 0″.6 from the main star, as was the case in 1795, which was the first occultation of a star by another star ever observed.

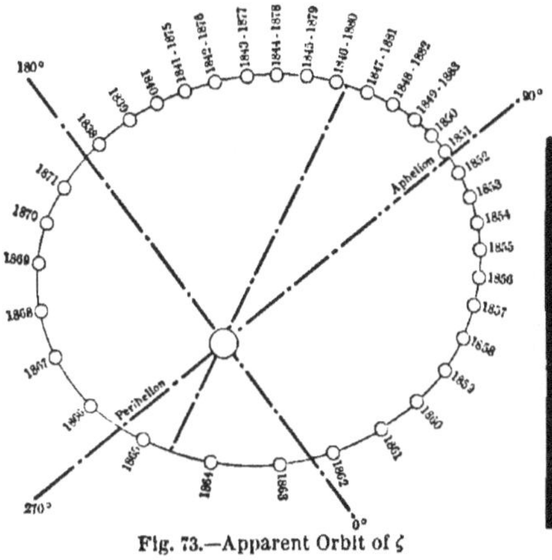

Fig. 73.—Apparent Orbit of ζ

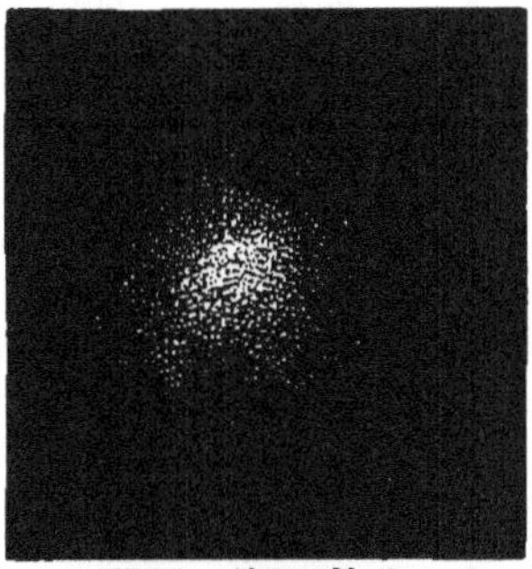

Fig. 74.—Cluster M. 13.

M. 13—Is one of the finest clusters, 8′ in diameter, and is composed of more than 5,000 stars from the 10th to the 15th magnitude; visible to the naked eye on clear nights, when there is no moon; it is about one-third the distance between η (eta) and ζ (zeta).

M. 92—Is another nice cluster of about the same diameter, but not so easy; see between 31,312 and ι (iota).

68 u—Varies from the 4th to the 6th magnitude in 40 days; it is a double star.

70—Double, most likely variable; noted 5th magnitude by Hevelius, in 1660; of 4th and 5th by Piazzi; of 6th by Argelander; three times of 4th magnitude by Lalande; of 5½th by Flammarion. Piazzi saw it twice double, but generally single; C. Mayer saw it double; W. Herschel found it single; Sir James South noted at 3′ 38″ a companion of 9th magnitude one time, of 10th another time, and of 11th still another time. Flammarion noted it of 9th magnitude in 1882 (Les Etoiles, page 225). It is a little south of the line between δ (delta) and λ (lambda), nearer to the first one.

Mr. Belopolsky, in 1888, obtained for η (*eta*) a parallax=0″.40±0″.072; and for π (*pi*) a parallax=0″.11±0″.063.

The sun, planets, moon, earth—our entire system—are now going toward a part of this constellation, marked on our planisphere. Are we going directly in that direction? are we describing an orbit around some other star? The future will decide.

AQUILA AND ANTINOUS.

Aquila is, according to some authors, the eagle of Jupiter, and according to others, Merops, which was changed into an eagle and placed among the stars by Juno. It is an old constellation already mentioned by Eudoxus.

Antinous was a young man of great beauty who was drowned in the Nile 132 years A. D. It was introduced as a constellation during the reign of the Emperor Adrian, who loved him so much that he erected temples in his honor and gave his name to a city.

DESIGNA-TION.	MAGNI-TUDE.	POSITION		DESIGNA-TION.	MAGNI-TUDE.	POSITION	
		R. A. h. m.	1880 DECL. ° ′			R. A. h. m.	1880 DECL. ° ′
α dbl.	1.5	19.45	+ 8.33	15 *h*, dbl.	5.7	18.59	− 4.13
β dbl.	4.0	19.49	+ 6. 6	4	5.5	18.39	+ 1.56
γ dbl.	3.3 red	19.41	+10.19	5 trip.	6.0	18.40	− 1. 5
δ dbl.	3.4	19.19	+ 2.52	11 dbl.	5.5	18.54	+13.37
ε dbl.	4.1	18.54	+14.54	12	4.0 yel.	18.55	− 5.58
ζ dbl.	3.0	19. 0	+13.41	18	5.5	19. 1	+10.53
η	3.V.	19.46	+ 0.42	19	5.8	19. 3	+ 5.53
θ	3.0	20.05	− 1.11	20	5.9	19. 6	− 8. 9
ι	4.4	19.30	− 1.33	21 dbl.	5.7	19. 8	+ 2. 6
κ	5.4	19.30	− 7.18	23 dbl.	5.7	19.12	+ 0.50
λ	3.3	19. 0	− 5. 4	51	5.8	19.44	−11. 5
μ	4.V.	19.28	+ 7. 7	56 dbl.	6.2	19.48	− 8.54
ν	5.4	19.20	+ 0. 6	57 dbl.	6.4	19.48	− 8.33
ξ dbl.	5.2	19.48	+ 8. 9	66	5.8	20. 7	− 1.22
ο	5.7	19.23	+ 1.43	69	5.4	20.23	− 3.17
π dbl.	6.0	19.43	+11.31	70	5.2 red	20.30	− 2.58
ρ	5.5	20. 9	+14.50	71	4.6	20.32	− 1.31
σ dbl.	5.7	19.33	+ 5. 7	R	7.V. red	19. 1	+ 8. 3
τ	5.9	19.58	+ 6.56	*	7.0 red	19. 4	+23.59
υ	6.2	19.40	+ 7.20	*	6.3 red	19.10	+18.19
φ	5.5	19.50	+11. 6	*	6.4 org.	18.41	+18.35
χ	5.8	19.37	+11.32	*	5.9 org.	18.51	+17.58
ψ	6.4	19.39	+13. 1	*	6.5 org.	18.55	+22.39
ω	6.0	19.12	+11.23	*	6.5 org.	18.57	+ 8.12
				*	5.0 org.	19.21	+19.34
28 A, dbl.	6.0	19.14	+12.10	*	6.2 org.	19.21	+19.39
31 *b*	5.8	19.19	+11.41	*	6.9 org.	19.24	+ 2.39
35 *c*	6.0	19.23	+ 1.42	*	7.1 org.	19.25	+ 1.46
27 *d*	5.9	19.14	− 1. 7	*	7.2 org.	19.27	+ 4.46
36 *e*	5.6 red	19.24	− 3. 3	*	6.9 org.	19.28	+ 5.12
26 *f*	5.7	19.14	− 5.39	*	7.V. org.	20.20	+ 9.40
14 *g*	5.8	18.57	− 3.52	M. 11	cl.	18.45	− 6.27

NOTES.

α (*Alpha*) *Altair*—Double; magnitudes 1.7 and 10; distance, 2′ 36″; difficult on account of the difference of magnitude. Struve gave for its parallax 0″.181±0″.094 and Mr. Elkin, of Yale College, in 1887, 0″.199±0″.047; and in taking the average, 0″.19, it put the distance at 1,086,000 times the distance of the earth from the sun, or 100 trillions of miles, and it takes the light a little over 17 years to reach us (Revue d'Ast., 1889; page 450).

η (*Eta*)—Varies from 3.5 to 4.7 in 7 days 4 hours 13 minutes and 53 seconds regularly; this remarkable variation can be followed by the naked eye; very interesting observation.

15 *h*—Double; magnitudes 5.7 and 7.5; distance, 35"; very easy pair.

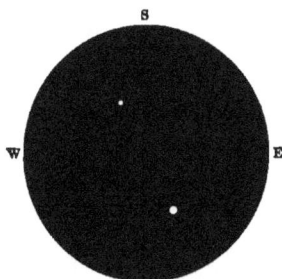

Fig. 75.—Double Star 15 *h*.

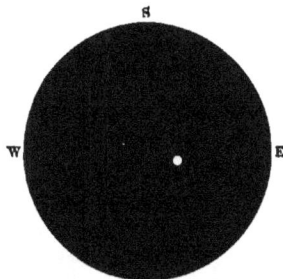

Fig. 76.—Double Star 11.

57—Double; magnitudes 6.4 and 7.0; distance, 35"; the two stars are sometimes of the same color and sometimes different.

11—Double; magnitudes 5.5 and 9; distance, 17"; pair in rapid motion.

23—Double; magnitudes 5.7 and 10; distance, 3"; delicate pair; the companion has its brilliancy developed more than in ordinary cases by using stronger power.

γ (*Gamma*) *Tarazed*—Is also a difficult double in a rich field; use small power or an opera glass.

β (*Beta*) *Alshain*—Is also double.

M. 11—Is a cluster with several small double stars in the field (see between 12 and 6 of Scutum Sobiesii).

μ (*Mu*)—Varies from the 4th to the 6th magnitude. Many of the stars of this constellation are remarkable for their variability, very few catalogues giving the same magnitudes from one century to another.

Fig. 77.—Cluster Messier 11.

SCUTUM SOBIESII.

Scutum Sobiesii, or Sobieski's Shield, is a constellation introduced by Hevelius about 1660, in honor of the Polish hero Sobieski.

DESIGNA- TION.	MAGNI- TUDE.	POSITION		DESIGNA- TION.	MAGNI- TUDE.	POSITION	
		R. A. h. m.	1880 DECL. ° '			R. A. h. m.	1880 DECL. ° '
1	3.8	18.29	− 8.20	*	6.5 org.	18.26	−14.57
2 dbl.	5.2	18.36	− 9.10	*	7.0 org.	18.38	− 6.39
3	5.3	18.37	− 8.24	M. 16	cl.	18.12	−13.51
6 —	4.6	18.42	− 4.52	M. 17	neb.	18.14	−16.14
9 —	5.5	18.51	− 6. 1	M. 18	cl.	18.13	−17.12
R	5.V.	18.41	− 5.50	M. 24	cl.	18.11	−18.28
34113	4.8	18.22	−14.39				

NOTES.

This constellation, forming part of Aquila and Antinous, is not recognized by many authors; in R. A. Proctor's new atlas, for instance, it does not appear. It is very rich in nebulæ, and the Milky Way is very bright in it. It is in that region that William Herschel counted 331,000 stars in five square degrees of it. With an opera glass or a telescope with large field and small power it is beautiful and astonishing. (See Fig. 78.)

SOUTH

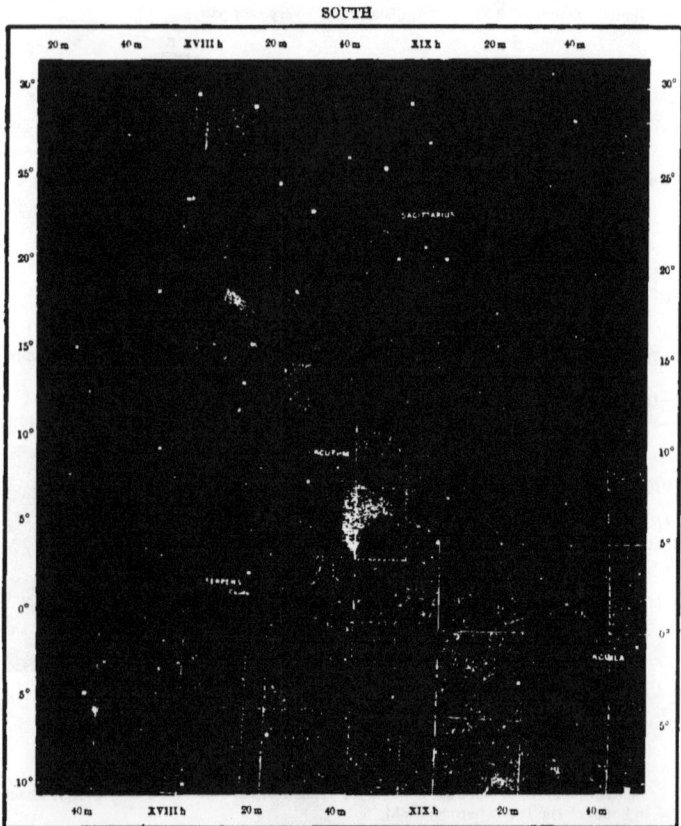

Fig. 78.—The Milky Way in Scutum Sobiesii and adjoining Constellations.

M. 17—The first one below 34,113 in our planisphere is the famous nebulæ, horseshoe-shaped; it is one of the most interesting; it has two centers of concentration, and in comparing the sketches made at different periods by different observers the representations vary very much; is it a world in formation?(Flam., Les Etoiles, pages 237-238.)

M. 18 and M. 24—Are two clusters also indicated below the above in our planisphere.

R.—Varies from the 5.2 to 7.3 in 71 days and 2 hours.

Fig. 79.—Nebula Messier 17.

OPHIUCHUS.

Ophiuchus, or Serpent Bearer, represents the god of medicine, Æsculapius; as the serpent is a symbol of prudence, so it was one of his attributes. This constellation is one of the forty-eight of the ancients, and appears already during Eudoxus' time.

DESIGNA- TION.	MAGNI- TUDE.	POSITION R. A. h. m.	1880	DECL. ° '	DESIGNA- TION.	MAGNI- TUDE.	POSITION R. A. h. m.	1880	DECL. ° '
α dbl.	2.0	17.29		+12.39	53 f, dbl.	6.0	17.29		+ 9.40
β	3.0 yel.	17.38		+ 4.37	20	5.0	16.43		−10.34
γ	3.8	17.42		+ 2.45	30	5.5 org.	16.55		− 4. 3
δ dbl.	3.1 red	16. 8		− 2.23	39 dbl.	5.8	17.11		−24. 9
ε	3.4 yel.	16.12		− 4.23	41	5.1	17.10		− 0.19
ζ	3.0	16.31		−10.19	P. XVII, 99	4.9	17.20		− 4.59
η dbl.	2.7	17. 3		−15.34	58	5.4	17.36		−21.37
θ	3.7	17.15		−24.53	66	5.2	17.54		+ 4.23
ι	4.4	16.48		+10.22	67 dbl.	4.5	17.55		+ 2.56
κ	3.4 yel.	16.52		+ 9.34	68 dbl.	4.7	17.56		+ 1.19
λ bin.	3.8	16.25		+ 2.15	70 bin.	4.4	17.59		+ 2.32
μ	4.7	17.31		− 8. 3	71	7.0	18. 2		+ 8.43
ν	3.6	17.52		− 9.45	72	3.6	18. 2		+ 9.33
ξ	5.0	17.14		−20.58	74 dbl.	5.5	18.15		+ 3.19
ρ dbl.	5.0	16.18		−23.10	R	8.V. org.	17. 1		−15.56
σ	4.9	17.21		+ 4 15	S	8.V.	16.27		−16.54
τ bin.	5.2	17.56		−− 8.11	Temp. of 1604	....	17.23		−21.23
υ	5.3	16.21		− 8. 6	Temp. of 1848	....	16.53		−12.42
φ	4.6	16.24		−16.21	*	8.0 red	17.38		−18.36
χ	4.7	16.20		−18.11	*	6.4 org.	16. 3		+ 8.52
ψ	4.8	16.17		−19.45	*	7.3 org.	17.52		+ 2.44
ω	4.7	16.25		−21.12	*	6.0 rose	17.16		−28. 2
					M. 14	cl.	17.31		− 3.10
36 A, bin.	5.5	17. 8		− 26.25	J. H. 1992	cl.	17.56		+11. 2
44 b	4.7	17.19		−24. 4	M. 23	cl.	17.50		+18.59
55 c	5.5	17.24		−23.52	M. 10	cl.	16.51		+ 3.55
45 d	4.6	17.20		−29.45	M. 12	cl.	16.41		+ 1.44
e	5.7	17.13		+11. 0					

NOTES.

α (Alpha)—Is also called Ras Alhague; β (beta) Cebalrai.

T (1604)—Observed by Fabricius and Kepler; it was brighter than Jupiter the day of its apparition, the 10th of October, 1604—nearly as bright as Venus; in January, 1605, it was still brighter than Antares; in February it was of the 2d magnitude; in March of the 3d magnitude, but as this part of the constellation, after that date, was lower than the horizon it could not be followed any longer; six months later it could not be found with the naked eye and, unfortunately, the telescope was not invented. Its position was not well defined, but must have been between ξ (zi) and 58; there is no star above the 9th magnitude in this region (Flam., Les Etoiles, page 248).

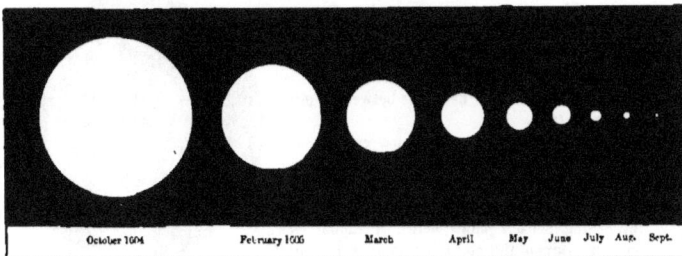

Fig. 80.—Diagram showing the variations of the Temporary of 1604.

T (1848)—Between η (*eta*) and 20 another temporary was observed by Hind, the 28th of April, 1848; it was noted 4½ magnitude; it was visible to the naked eye until May 11th and then went down to the 6th magnitude; in July it was of the 7th; in June, 1849, it was seen of the 10th; since 1850 it is of the 11th magnitude.

R., S. and T.—Are also variables but require strong power to see them.

A.—Double; magnitudes 4.5 and 6; distance, 4".3; as the first varies from the 4th to the 6th magnitude, the components are sometimes equal (such was the case in 1877). At 14' there is a star of the 7th magnitude, traveling in the same direction (S. S. W.) as the above, with a velocity of 1".27 per year; these three stars form a physical system.

70—Binary; magnitudes 4.4 and 6; distance, 2".9 in 1880; **revolution about 90 years.** If the parallax, 0".168 is correct, the two suns are revolving around each other at an average distance of 2,680 millions of miles apart, which is a little less than the distance from Neptune to the sun; if the star 70 was the same weight as our own sun, the companion would revolve in a little less than 164 years, but as it revolves in 90 years it proves that the star 70 weighs nearly three times as much as the sun; and as the sun weighs 324,000 times as much as the earth the conclusion is that 70 of Ophiuchus weighs 985,000 times as much as the earth (Flam., Les Etoiles, page 253). This remarkable pair has, besides, a proper motion of 1".1 per year, which represents a minimum of 350 millions of miles.

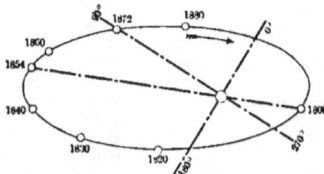

Fig. 81.—Apparent Orbit of Double Star 70.

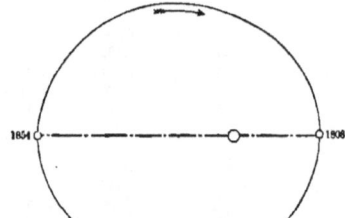

Fig. 82.—Real Orbit of 70.

67—Double; magnitudes 4.5 and 8; distance, 55"; in the same field there is a little orange star of the 7th magnitude; very easy. Mr. Burnham found a star of the 15th magnitude at a distance of 6".7 in 1889.

39—Double; magnitudes 5.7 and 7.5; distance, 12"; nice pair; yellow and blue.

ρ (*Rho*)—Double; magnitudes 5.0 and 7.5; distance, 3".8; delicate pair; yellow and blue.

τ (*Tau*)—Binary; magnitudes 5.2 and 6; distance, 1".8 in 1886; period of revolution about 218 years.

λ (*Lambda*)—Binary; magnitudes 3.8 and 6; distance, 1".5 in 1880; revolution, 233 years.

Σ 2173—Binary; magnitudes 6 and 6; distance, 0".8; revolution, 45 years; close pair.

M. 14—Is a cluster in a rich field; see between γ (*gamma*) and μ (*mu*).

Between α (*alpha*) and 72 there is another rich cluster, very bright, almost visible to the naked eye; use opera glass.

This constellation has several other nebulæ more or less interesting (see our planisphere). Poczobut took out some stars of this constellation to form Taurus Poniatowskii (Poniatowski's bull).

Fig. 83.—Cluster Messier 14.

SERPENS.

This constellation is one of the forty-eight constellations of the ancients (see Ophiuchus).

DESIGNA-TION.	MAGNI-TUDE.	POSITION		DESIGNA-TION.	MAGNI-TUDE.	POSITION	
		R. A. 1890 DECL.				R. A. 1890 DECL.	
		h. m. ° ′				h. m. ° ′	
α dbl.	2.6	15.38	+ 6.48	ν	6.0	15.42	+14.30
β dbl.	3.3	15.41	+15.48	φ	6.0	15.52	+14.46
γ	3.8	15.51	+16. 3	χ	5.8	15.36	+13.14
δ bin.	3.3	15.29	+10.57	ψ	6.2	15.38	+ 2.54
ε	3.7	15.45	+ 4.51	ω	5.7	15.44	+ 2.34
ζ	4.8	17.54	− 3.41				
η dbl.	3.4	18.15	− 2.56	11 A¹	6.0	15.27	− 0.47
θ dbl.	4.4	18.50	+ 4. 3	25 A²	5.8	15.40	− 1.26
ι	4.9	15.36	+20. 3	36 b	5.6	15.45	− 2.43
κ	4.0 org.	15.43	+18.31	60 c	5.9	18.23	− 2. 4
λ	4.7	15.41	+ 7.43	59 d, dbl.	5.6	18.21	+ 0. 7
μ	3.3	15.43	− 3. 4	e	6.1	18.32	− 0.25
ν dbl.	4.6	17.14	−12.43	5 dbl.	5.2	15.13	+ 2.14
ξ	3.7	17.31	−15.19	R	6.V. org.	15.45	+15.30
ο	4.7	17.35	−12.40	S	8.V. red	15.16	+14.45
π	4.7	15.57	+23. 9	*	6.7 red	15.31	+15.30
ρ	4.8 red	15.46	+21.21	M. 5	cl.	15.12	+ 2.32
σ	5.4	16.16	+ 1.19	H. VIII, 72	cl.	18.22	+ 6.29
τ¹	5.5	15 20	+15.51	J. II. 1929	cl.	15.31	+ 6.25

NOTES.

α (*Alpha*) *Unukalhai*—Is a double star.

θ (*Theta*)—Double; magnitudes 4.4 and 5.0; distance, 21″; easy pair.

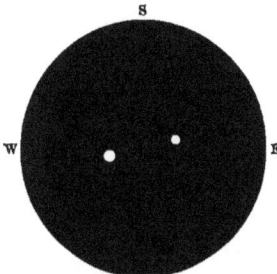

Fig. 83.—Double Star θ

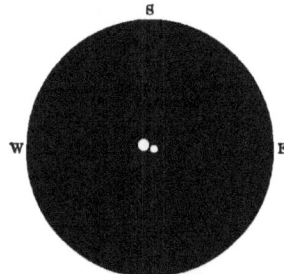

Fig. 84.—Double Star δ

δ (*Delta*)—Binary; magnitudes 3.4 and 5.0; distance, in 1880, 3″.5; the companion is a variable and the two stars have often been seen of same magnitude; the orbital motion is quite slow, only 49° in 98 years; the entire revolution must take about 900 years (Flam., Les Etoiles, page 257).

ν (*Nu*)—Double; magnitudes 4.6 and 9; distance, 51″; easy pair.

5—Double; magnitudes 5.2 and 10; distance, 10″.

M. 5—Is a rich cluster. W. Herschel counted 200 stars in it, but in the center they are so close together that it is nearly impossible to determine the number (see near 5).

H. VIII, 72—Is also a cluster, almost visible to the naked eye (see with an opera glass between θ (*theta*) and 72 of Ophiuchus). An opera glass will also show a little cluster N. E. of α (*alpha*).

Fig. 85.—Cluster, Messier 5.

ZODIACAL CONSTELLATIONS.

PISCES. 🐟

This constellation appeared in the astronomical sphere of Eudoxus; some said it was called the Fishes because it appears at the time of the rainy days; others said that they were consecrated to Venus and Cupid, who were changed into fishes to escape the pursuit of the Giants; for this reason the Syrians never ate fish, for in so doing they feared they might eat one of their gods. It is now the first constellation of the Zodiac.

DESIGNA-TION.	MAGNI-TUDE.	POSITION R. A. 1880 h. m.	POSITION DECL. ° ′	DESIGNA-TION.	MAGNI-TUDE.	POSITION R. A. 1880 h. m.	POSITION DECL. ° ′
a bin.	4.0	1.56	+ 2.11	5 A	5.6	23. 3	+1.28
β	4.5	22.58	3.10	7 b	5.5	23.14	4.44
γ	3.8	23.11	2.38	31 c¹	6.3	23.56	8.17
δ	4.5	0.42	6.56	32 c²	5.8	23.56	7.49
ε	4.3	0.57	7.15	41 d	5.3	0.14	7.32
ζ dbl.	4.9	1. 7	6.57	80 e	5.6	1. 2	5. 1
η	3.6	1.25	14.44	89 f	5.2	1.12	2.59
θ	5.4	23.22	5.44	82 g	5.5	1. 4	30.47
ι dbl.	4.2	23.34	4.59	68 h	6.0	0.51	28.21
κ	4.8	23.21	0.36	65 i, dbl.	6.0	0.43	27. 4
λ	4.7	23.36	1. 7	67 k	6.0	0.49	26.34
μ dbl.	5.0	1.24	5.32	91 l	5.5	1.14	28. 7
ν	4.6 yel.	1.35	4.53	19	5.0 org.	23.40	+ 2.49
ξ	4.7	1.47	2.36	27	5.2	23.53	− 4.13
ο	4.4	1.39	8.33	29	5.0	23.56	− 3.42
π trip.	5.8	1.31	11.32	30	4.5	23.56	− 6.41
ρ	5.6	1.29	18.35	33	4.9	23.59	− 6.23
σ	5.5	0.56	31.10	35 dbl.	6.0	0. 9	+ 8.10
τ	5.4	1. 5	29.27	51 dbl.	6.0	0.26	6.17
υ	4.4	1.13	26.38	55 dbl.	5.8	0.34	20.47
φ dbl.	4.8	1. 7	23.57	58	5.4	0.41	11.20
χ	4.8	1. 5	20.24	77 dbl.	6.0	1. 0	4.16
ψ¹ dbl.	4.9	0.59	20.50	94	5.3	1.21	18.39
ψ²	5.8	1. 1	20. 6	100 dbl.	7.0	1.29	11.57
ψ³	6.0	1. 3	19. 1	R	8.V. red	1.24	2.16
ω	4.2	23.53	6.12	•	7.0 org.	1.10	25. 8

NOTES.

a (*Alpha*)—Binary; magnitudes 4 and 5; distance, 3″.1 in 1880; very slow orbital motion, only 14° in 100 years; revolution, 2,570 years.

ζ (*Zeta*)—Double; magnitudes 4.9 and 6; distance, 24″; easy and bright pair.

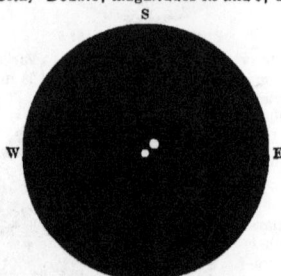

Fig. 86.—Double Star a

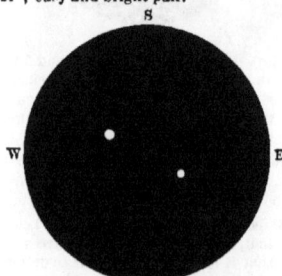

Fig. 87.—Double Star ζ

ψ (*Psi*)—Double; magnitudes 5.4 and 5.4; distance, 30″; very easy pair, in motion.
77—Double; magnitudes 6 and 7; distance, 33″; very easy pair.
65 ξ—Double; magnitudes 6 and 8; distance, 4″.5; the components are sometimes of the same magnitude; it is a delicate pair.
35—Double; magnitudes 6 and 8; distance, 12″; easy pair.
51—Double; magnitudes 6 and 9; distance, 28″; small companion but easy pair; pearl-white and pale lilac.
55—Double; magnitudes 6 and 9; distance, 6″; pretty pair; orange and sapphire—nice contrast.
This constellation has also several variables, but they require a good instrument.

ARIES.

Aries, or the Ram, is a very old constellation; it was called Jupiter Ammon and Chrysommallus, or the Golden Fleece; Eudoxus had it on his sphere, and in all probability found it already established by the Egyptians.

DESIGNA-TION.	MAGNI-TUDE.	POSITION		DESIGNA-TION.	MAGNI-TUDE.	POSITION	
		R. A. 1880	DECL.			R. A. 1880	DECL.
		h. m.	° ′			h. m.	° ′
α (Hamal) dbl.	2.2 yel.	2. 0	+22.54	46 ρ	6.0	2.50	+17.32
β dbl.	3.0	1.48	20.13	σ	5.8	2.45	14.35
γ dbl.	3.9	1.47	18.42	61 τ¹	5.0	3.14	20.43
δ	4.1	3. 5	19.16	63 τ²	5.5	3.16	20.18
ε bin.	4.8	2.52	20.52	1 dbl.	6.0	1.44	21.41
ζ	4.9	3. 8	20.35	7	6.5	1.49	22.59
η	5.5	2. 6	20.39	14 trip.	5.4	2. 2	25.23
θ	5.7	2.11	19.20	30 dbl.	6.0	2.30	24. 7
ι	5.8	1.51	17.14	33 dbl.	5.8	2.34	26.32
κ	5.7	2. 0	22. 4	35 dbl.	5.0	2.36	27.13
λ dbl.	5.3	1.51	23. 1	38	5.0	2.38	11.56
μ dbl.	5.8	2.36	19.30	39	4.9	2.41	28.44
ν	6.0	2.32	21.26	41 quad.	3.8	2.43	26.46
ξ	5.5	2.18	10. 4	P. III, 32	5.2	3.13	28.32
ο	6.0	2.38	14.48	R	8.V. org.	2. 9	24.30
π trip.	5.6	2.42	16.57	T	8.V. yel.	2.42	17. 1

NOTES.

α (*Alpha*)—Also called *Hamal*, is a double star. β (*Beta*) *Sheratan* is also double.
γ (*Gamma*) *Mesartim*—Double; magnitudes 4.2 and 4.5; distance, 8″.9; very easy pair; it is the first star noted as double. In 1664 Hooke, contemporary of Newton, In looking at the comet of that year discovered it and said: "*I took notice that it consisted of two small stars very near together; a like instance to which I have not else met with in all the heaven.*" Since that day many double stars have been discovered, and every day adds to the list.

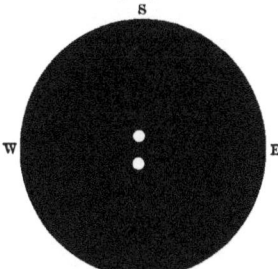

Fig. 88.—Double Star γ

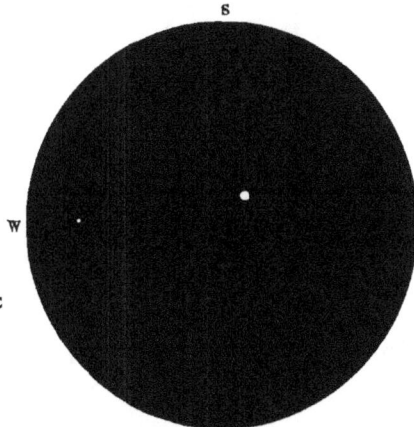

Fig. 89.—Triple Star 14.

14—Triple; magnitudes 5.4-10 and 9; distances, 82″ and 106″; white, blue and lilac.

30—Double; magnitudes 6.0 and 7.0; distance, 38″; very easy pair; this region is rich in double stars.

λ (*Lambda*)—Double; magnitudes 5.3 and 8; distance, 38″; very easy pair.

π (*Pi*)—Triple; magnitudes 5.6-8.5 and 11; distances, 3″ and 25″; somewhat difficult.

33—Double; magnitudes 5.8 and 9; distance, 28″.

ε (*Epsilon*)—Binary; magnitudes 5 and 6; distance, 1″.3 in 1880; very close pair; in 1800 the two stars were at one-tenth of a second apart and most likely eclipsed.

This constellation was, 2,000 years ago, the first one of the Zodiac.

The stars 33, 35, 39 and 41 were formed in a little constellation by Bartschius, son-in-law of Kepler, and called Musca—*the Fly*—on his *Celestial Globe*, in 1623.

TAURUS. ♉

Taurus, or the Bull, is one of the old Egyptian constellations; it represented the bull Apis; the Greeks identify it with the bull which carried off Europa. It was already mentioned by Homer and Hesiod.

The Pleiades and the Hyades are noted in Job (xxxviii, 31); also by Homer and Hesiod. Aldebaran is the bull's eye, and the Hebrews called it "Aleph " (God's Eye).

DESIGNA-TION.	MAGNI-TUDE.	POSITION			DESIGNA-TION.	MAGNI-TUDE.	POSITION		
		R. A. h. m.	1880	DECL. ° ′			R. A. h. m.	1880	DECL. ° ′
α (Aldebaran) trip.	1.4 red	4.29		+16.16	*g*	6.2	3. 6		+6.13
β dbl.	2.0	5.19		28.30	57 *h*	6.0	4.13		13.45
γ dbl.	4.1	4.13		15.20	97 *i*	5.7	4.44		18.38
δ¹	4.0	4.16		17.15	98 *k*	6.0	4.51		24.52
δ²	5.9	4.17		17.10	106 *l*	5.8	5. 1		20.16
ε	3.7	4.22		18.55	104 *m*	5.5	5. 0		18.29
ζ	3.5	5.30		21. 5	109 *n*	5.9	5.12		21.57
η (Pleiades)	3.0	3.40		23.44	114 *o*	6.0	5.20		21.50
θ¹	3.9	4.22		15.42	44 *p*	6.2	4. 3		26.10
θ²	4.2	4.22		15.36	66 *r*	5.4	4.17		9.11
ι	5.0	4.57		21.25	4 *s*	5.5	3.24		10.56
κ¹	4.8	4.18		22. 1	6 *t*	6.0	3.26		8.58
κ²	6.5	4.18		21.55	29 *u*	5.7	3.39		5.40
λ	3.V.	3.54		12. 9	10	4.5	3.31		0.02
μ	4.4	4. 9		8.35	40	5.4	3.57		5.06
ν	3.9	3.57		5.39	41	5.4	3.59		27.16
ξ	3.5	3.21		9.19	47	5.2	4. 7		8.58
ο	3.4	3.18		8.36	48	6.V.	4. 9		15. 6
π	5.8	4.20		14.26	68	5.0	4.18		17.40
ρ	5.6	4.27		14.35	105	6.0	5. 1		21.32
σ¹	5.4	4.32		15.32	119	5.6 red	5.25		18.30
σ²	5.4	4.32		15.40	121	5.8	5.28		23.58
τ dbl.	4.5	4.35		22.43	125	6.0'	5.32		25.50
υ¹	4.8	4.19		22.32	126 ·	5.9 ·	5.34		16.29
υ²	6.0	4.20		22.43	132	5.7	5.51		24.32
φ dbl.	5.5	4.13		27. 4	133 trip.	5.5	5.41		13.52
χ dbl.	5.7	4.15		25.20	134	5.4	5.42		12.38
ψ	5.6	4. 0		28.40	136	5.6	5.45		27.35
ω¹	5.8	4. 2		19.17	139	5.7	5.50		25.56
ω²	6.2	4.10		20.17	P. VI, 99	4.9	4.23		20.56
					P. IV, 246	5.3	4.50		16.58
37 A¹	4.9	3.58		21.45	Σ 730 dbl.	6.0	5.25		17. 0
39 A² dbl.	6.4	3.58		21.39	R	8.V. org.	4.22		9.54
79 *b*	5.8	4.22		12.46	U	8.V.	4.15		19.32
90 *c*¹	4.4	4.31		12.16	*	7.V.	3.47		7.25
93 *c*²	5.5	4.33		11.57	*	7.7 red	5.39		20.38
88 *d* dbl.	4.6	4.29		9.54	*	8.5 red	5.38		24.22
30 *e* dbl.	5.0	3.42		10.46	*	6.5 org.	4.15		20.32
5 *f*	4.7	3.24		12.32	M. 1	neb.	5.27		21.56

NOTES.

This constellation was the first one of the Zodiac during the Egyptian period, the sun crossing the ecliptic near Aldebaran, 3,000 years ago, at the vernal equinox.

Pleiades—This beautiful cluster contains over 1,000 stars, six or seven visible to the naked eye; it was named from the daughters of Atlas and the nymph Pleione. In our diagram we show the 14 principal stars, as seen with an opera glass; also, the names of the entire family. Alcyone is η (*cta*) of Taurus. The position and the magnitudes of these stars are very difficult to be identified from the catalogues and maps published at different times, and some changes have been noted from the first observations taken over 2,000 years ago and at the present time; most of them are variables and only a few years of observations indicate some difference in their magnitude.

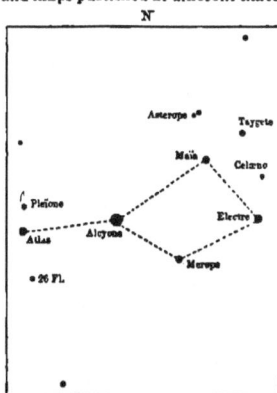

NAMES.	MAGNI-TUDE.	POSITION		
		R. A.	1880	DECL.
		h. m. s.		° ′
Celæno........	6.5	3.37.40		20.54,7
Electra........	4.5	3.37.45		23.44,1
Taygete.......	5.8	3.38. 4		24. 5.4
Maia...........	5.0	3.38.41		23.59,8
Asterope1 } ...	6.5 {	3.38.45		24.10,7
Asterope2 }	7.0 {			
Merope	5.5	3.39.12		23.34,4
Alcyone	3.0	3.40.21		23.44,0
Atlas..........	4.6	3.42. 2		23.41,2
Pleione	6.3	3.42. 3		23.46,2

Fig. 90.—The 14 principal Pleiades.

As our object is to describe the heavens as they are now, we have given the position, the names and the magnitude of the Pleiades for the year 1880; the amateurs who wish to go deeper with the study of this interesting group can refer to M. Flammarion's book, "Les Etoiles," pages 289 to 306.

Messrs. Paul and Prosper Henry, after taking the photograph of the Pleiades the 16th of December, 1885, discovered a bright nebula starting from Maïa, going a little west, then turning abruptly toward the north and measuring about 3 minutes; they photographed the same region again on the 16th of November, the 8th and 9th of December, 1885, and the 8th of January, 1886, and the nebula was visible on the plates each time; still, in looking through all the telescopes of the Paris Observatory they could not see it; it was identified later at Pulkova Observatory (Revue d'Ast., 1886; page 45).

In 1859, Mr. Tempel, of Marseilles, discovered at Venice, the 23d of October, 1859, a nebula starting from Merope, which he thought was a comet; he saw it again at Marseilles several times in 1860. Mr. Julius Schmidt, of Athens, never noticed it from 1844 until the 5th of February, 1861; the atmosphere that night was extraordinarily pure. Mr. Chacornac observed it at Paris the 16th of September, 1862, and Mr. Webb in 1863 and 1865; it was then very faint and some experienced observers, such as Burnham, never saw it; in November, 1890, Mr. Barnard also saw it, and since then it has been photographed several times. This nebula is very likely variable.

The Henry Brothers took another photograph of the same region the 16th of November, 1887 (Fig. 90), showing several new nebulæ, each one in the neighborhood of the principal stars and connected by faint nebulosities, one of them especially very straight, measuring 37 minutes and passing through seven small stars nearly parallel to the direction of Pleione to Alcyone, starting from a star about half way between Alcyone and Maia and ending a little N. W. of Pleione. This photograph was taken again the 14th of December, 1887, and made over twice, showing always the same details, therefore they must really exist. The Pleiades show no sign of a parallax, but supposing that they are as near to us as 61 Cygni or Aldebaran and offer a parallax of 0″.5, the line above described would represent 410 billions of miles and in the same proportion would be 650 millions of miles in width. These figures are certainly the minima (1888, Revue d'Ast., pages 401 and 404).

NORTH

Fig. 91.—The Pleiades, from a photograph taken by Messrs. Paul and Prosper Henry, in 1887.

The distance between Alcyone and Electra is 37 minutes, the distance from Merope to Maia is 25 minutes; the full moon measures only 31 minutes, and in looking at that beautiful cluster the impression is that the moon would cover the entire group; still it could be inserted between Taygete and Merope without touching these stars. When the moon passes in front of the Pleiades and eclipses one after the other, it is wonderful and one can hardly believe his own eyes.

From observations by Bradley in 1755, by Bessel in 1825 and 1840, by Wolf in 1874, all the Pleiades have a proper motion a little east of the south; as our system goes in that direction and the effect may be due to our own motion, we reproduce M. Flammarion's diagram representing this remarkable illustration; the ends of the arrows show the position of each of the stars 10,000 years from now, if the movement is regular.

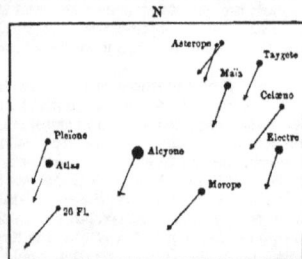

Fig. 92.—Proper motion of the Pleiades.

Hyades—Are another cluster, also visible to the naked eye; very nice with an opera glass; from observations made for 30 stars of this cluster it is evident that they all go in the same direction, S. E. S. toward Aldebaran, which has itself a motion in a direction inclined from 45° to 50° with the general motion of the Hyades.

α (*Alpha*) *Aldebaran*—Triple; magnitudes 1.4-14 and 11; distance, 30".4 and 115". The faint star is a difficult object on account of the brilliancy of Aldebaran and requires good power. Mr. Burnham found the nearest companion at Chicago in 1877, distance, 30".4, and in 1888, with the 36-inch telescope of Mt. Hamilton, discovered that the old companion was also double, distance 2". Aldebaran and Burnham companion have the same proper motion, 0".188 in the direction of 164°.4, and the old companion has an annual motion of 0".005 in the direction of about 109°.6. Fig. 93, made from a diagram by Mr. Burnham, shows the proper motion of Aldebaran and its companions.

Struve gave for the parallax of this star in 1854..............0".516±0".057
Elkin, of Yale College, in 1887-8, gave......................0".116±0".029
Hall, of Washington, in 1888...................... ....0".102±0".030

In taking the average of the three or 0".24 it would represent 874,000 times the distance of the earth from the sun, or 80 trillions of miles, and the light would have to travel more than 13 years and 9 months to reach us (Revue d'Ast., 1889; pages 445 and 450). It is one of the nearest stars to us.

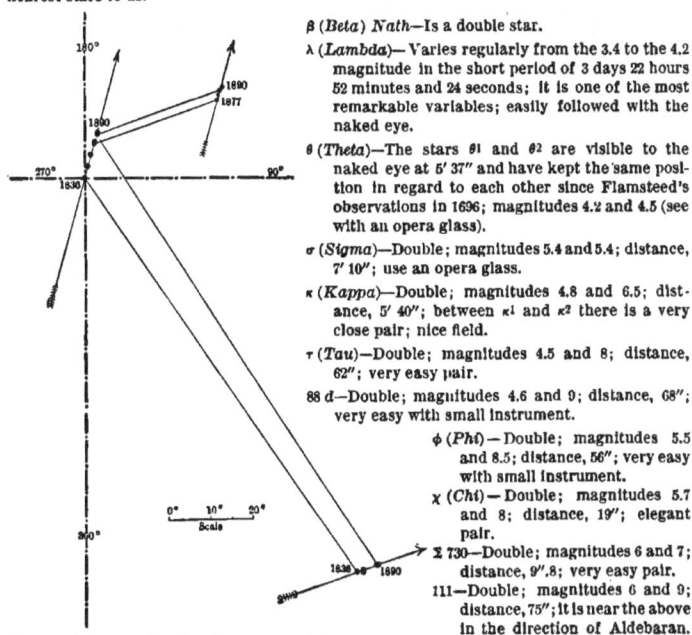

β (*Beta*) *Nath*—Is a double star.

λ (*Lambda*)— Varies regularly from the 3.4 to the 4.2 magnitude in the short period of 3 days 22 hours 52 minutes and 24 seconds; it is one of the most remarkable variables; easily followed with the naked eye.

θ (*Theta*)—The stars θ1 and θ2 are visible to the naked eye at 5' 37" and have kept the same position in regard to each other since Flamsteed's observations in 1696; magnitudes 4.2 and 4.5 (see with an opera glass).

σ (*Sigma*)—Double; magnitudes 5.4 and 5.4; distance, 7' 10"; use an opera glass.

κ (*Kappa*)—Double; magnitudes 4.8 and 6.5; distance, 5' 40"; between κ1 and κ2 there is a very close pair; nice field.

τ (*Tau*)—Double; magnitudes 4.5 and 8; distance, 62"; very easy pair.

88 d—Double; magnitudes 4.6 and 9; distance, 68"; very easy with small instrument.

φ (*Phi*)—Double; magnitudes 5.5 and 8.5; distance, 56"; very easy with small instrument.

χ (*Chi*)—Double; magnitudes 5.7 and 8; distance, 19"; elegant pair.

Σ 730—Double; magnitudes 6 and 7; distance, 9".8; very easy pair.

111—Double; magnitudes 6 and 9; distance, 75"; it is near the above in the direction of Aldebaran.

Fig. 93.—Diagram showing the proper motion of Aldebaran and its companions.

M. 1—Is the famous "Crab Nebula," see near ζ (*zeta*); discovered by Messier, and the first of his catalogue; he saw it by accident in looking at the comet of 1758; there are several small stars in it, but they do not belong to the nebula; a large telescope is necessary to show all the marvels of this beautiful object, which is no less than 3½ minutes wide by 5½ minutes long. (Fig. 95.)

This constellation contains several variables too small for common telescopes—we only note:

V.—Varies from the 8.6 to 12.8 magnitude in 170 days.

39 A²—East of A is a triple of magnitudes 6.4-9 and 9; distances, 26″ and 37″; the third star is coming nearer to the main star.

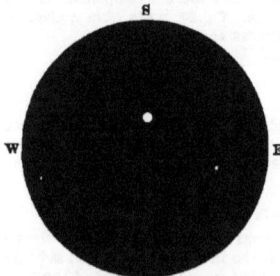

Fig. 94.—Triple Star 39 A²

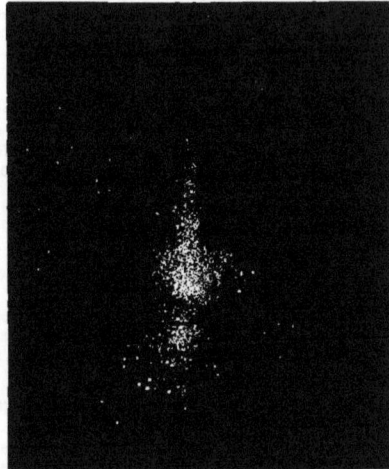

Fig. 95.—Nebula Messier 1.
As seen in Lord Rosse's telescope.

GEMINI. ♊

Gemini, or the Twins, according to Greek mythology, are Castor and Pollux, or Apollo and Hercules, all sons of Jupiter, thence the name "Dioscures" (Jupiter's Sons), sometimes given to them. Castor having been killed during the siege of Sparta, Pollux asked Jupiter to give half of his life to his brother so that they could live alternatively together. The king of the gods, to immortalize such proof of fraternal love, put them both among the constellations. According to others they are the inseparable gods Horus and Harpocrates of the Egyptians.

DESIGNA-TION.	MAGNI-TUDE.	POSITION		DESIGNA-TION.	MAGNI-TUDE.	POSITION	
		R. A. 1880 h. m.	DECL. ° ′			R. A. 1880 h. m.	DECL. ° ′
α (Castor) bin.	2 3	7.27	+32. 9	57 A	5.8	7.16	+25 17
β (Pollux) qdl.	1.9 org.	7.38	28.19	64 b¹	5.5	7.22	28.23
γ trip.	2.7	6.31	16.30	65 b²	5.0	7.22	28.11
δ dbl.	3.8	7.13	22.12	76 c	6.3	7.37	26. 5
ε dbl.	3.3 yel.	6.36	25.15	36 d	6.0	6.44	21.54
ζ trip.	4.V.	6.57	20.45	38 ε bin.	5.4	6.48	13.20
η	3.V.	6. 8	22.32	74 f	6.0	7.32	17.57
θ	4.V.	6.45	34. 6	81 g	5.8	7.39	18.49
ι	4.0	7.18	28. 3	1 (Propus)	5.0	5.57	23.17
κ dbl.	3.8	7.37	24.42	15 dbl.	6.0	6.21	20.52
λ dbl.	4.3	7.11	16.45	20 dbl.	6.0	6.25	17.52
μ dbl.	3.0 red	6.16	22.35	26	5.5	6.35	17.46
ν	4.6	6.22	20.17	30	5.7	6.37	13.21
ξ	3.9	6.38	13. 1	61 dbl.	6.V.	7.20	20.30
ο trip.	5.5	7.31	34.52	70 trip.	6.3	7.31	35.19
π trip.	5.7	7.40	33.44	85	6.0	7.48	20.13
ρ	4.6	7.21	32. 1	Σ 1083 dbl.	8.0	7.18	20.46
σ	4.5 org.	7.36	29.11	R	7.V. org.	7. 0	22.53
τ	4.8	7. 3	30.27	S	8.V. org.	7.36	23.44
υ	4.4 red	7.28	27.10	T	8.V. org.	7.42	24. 2
φ	5.4	7.46	27. 6	*	7.2 red	7. 8	22.11
χ	5.3	7.56	28. 8	*	7.4 red	6. 3	26. 2
ψ	5.7	8. 6	30. 2	*	6.7 red	6. 5	22.56
ω	5.8	6.55	24 23	M. 35	cl.	6. 1	24.21
				H. IV, 45	neb.	7.22	21. 9

NOTES.

a (*Alpha*) *Castor*—Binary; magnitudes 2.5 and 3.0; distance, in 1880, 5″.6; one of the finest double stars; the orbital motion takes nearly 1,000 years; at 73″ distance from Castor there is a small star of the 9½th magnitude which seems to form part of a trinary system. Castor is moving at the rate of 28 miles per second in an opposite direction from us, or 890 millions of miles per year. Johnson, in 1854-5, found a parallax of 0″.198 ± 0″.140; which is too uncertain to fix the distance of this beautiful pair from us.

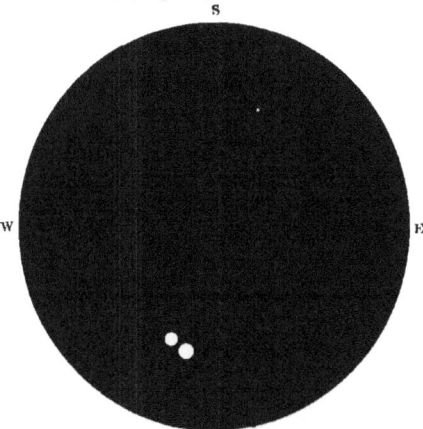

Fig. 96.—Castor and its small companion.

β (*Beta*) *Pollux*—Is a quadruple star, but requires good power, the companions being from the 10th to the 12th magnitude. This star is coming toward us with a velocity still greater than the one of Castor, about 40 miles per second, and the distance between the "two brothers" must be 6 trillions 250 billions of miles, farther apart now than at the time of Homer (Flam., Les Etoiles, page 319). Mr. Elkin, in 1887-8, found for its parallax 0″.068 ± 0″.047, but the uncertainty is too great to determine its distance (Revue d'Ast., 1889; page 448).

γ (*Gamma*) *Alhena*—Is a triple star.

ζ (*Zeta*) Varies from the 3.7 to the 4.5 magnitude in a period of 10 days 3 hours and 47 minutes; it has a companion of the 8th magnitude at a distance of 90″.

η (*Eta*)—Varies from the 3.2 to the 4.2 magnitude in a period of 230 days. It is a close double, discovered by Mr. Burnham in 1881; distance, 1″.

δ (*Delta*) *Wesat*—Double; magnitudes 3.8 and 8; distance, 7″; supposed to be binary.

ε (*Epsilon*) *Mebsuta*—Has a distant companion, 111″.

κ (*Kappa*)—Double; magnitudes 3.8 and 9; distance, 6″; orange and azure; the companion varies from the 8th to the 10th magnitude. Sir John Herschel thought that it was a planet of the main star and was receiving its light from it (Flam., Les Etoiles, page 320).

38 ι—Double; magnitudes 5.4 and 8; distance, 6″; the two stars are variables; the first one is gold-yellow, and the companion changes color from green to blue and from purple to red.

20—Double; magnitudes 6 and 7; distance, 20″; these two stars are also variables.

64—Double; magnitudes 6 and 9; distance, 60″.

M. 35—Near Propus is the famous cluster of Gemini, visible to the naked eye, composed of hundreds of stars from the 11th to the 12th magnitudes. A field of 19′ in diameter is entirely filled up with stars; splendid object.

H. IV, 45—Is a planetary nebula 2° S. E. of δ (*delta*); its spectrum shows the rays of hydrogen and nitrogen; it is absolutely gaseous. In the center there is a star of the 9th magnitude. This constellation contains also several variables too small for common telescopes.

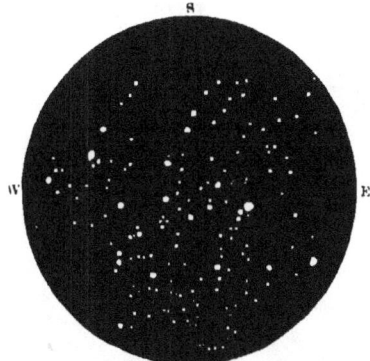

Fig. 97.—Cluster Messier 35.

CANCER. 🦀

Cancer, or the Crab, was known by Eudoxus, Hipparchus and Ptolemy, and its Greek name, " Karkinos," has the same meaning as the English name Crab.

It is claimed by some that the name of this constellation has been given it because the sun passed through it at the solstice of summer and then started to retrograde; but it was known more than 2,000 years ago, and this explanation is not applicable. According to others, it is the crab killed by Hercules when he was fighting against the Hydra.

DESIGNATION.	MAGNITUDE.	POSITION		DESIGNATION.	MAGNITUDE.	POSITION	
		R. A. 1880 h. m.	DECL. ° ′			R. A. 1880 h. m.	DECL. ° ′
a	4.2	8.52	+12.19	18 χ	5.6	8.13	−27.37
β	3.7	8.10	9.33	14 ψ	6.0	8.3	25.53
γ	4.4	8.36	21.54	2 ω¹	6.0	7.54	25.44
δ	4.3	8.38	18.35	4 ω²	6.3	7.54	25.25
ε (Præsepe)	cl.	8.33	19.58				
ζ trin.	4.8	8.5	18.1	45 A¹	5.5	8.37	13.6
η	5.6	8.26	20.51	50 A²	5.5	8.40	12.33
θ dbl.	5.5	8.25	18.31	49 b	6.0	8.38	10.31
ι dbl.	4.5	8.30	29.12	36 c	6.0	8.31	10.4
κ	5.0	9.1	11.9	20 d¹	6.0	8.16	18.43
λ	5.8	8.13	24.24	25 d²	6.3	8.19	17.27
9 μ¹	6.3	7.59	22.59	8	6.2	7.58	13.24
10 μ²	5.9	8.1	21.56	24 dbl.	6.7	8.20	24.55
ν	5.5	8.56	24.56	57 dbl.	6.0	8.47	31.2
ξ	5.0	9.2	22.33	60	5.8 org.	8.49	12.5
62 o¹	5.5	8.50	15.47	83	6.0	9.12	18.13
63 o²	6.0	8.51	16.2	P. VIII, 42	6.3	8.13	
82 π	5.8 red	9.8	15.27	R	7.V. org.	8.10	
55 ρ¹	6.0	8.45	28.49	S	8.V.	8.38	19..
58 ρ²	5.8	8.48	28.25	T	8.V. red	8.50	20.10
51 σ¹	6.0	8.46	32.58	V	7.V. org.	8.15	17.40
59 σ² dbl.	5.8	8.52	33.23	Σ 1311 dbl.	6.7	9.1	23.26
64 σ³	5.0	8.54	32.49	Σ 1177 dbl.	6.5	7.59	27.52
Σ 1298 dbl.	6.5	8.54	32.46	*	6.5 red	8.49	17.41
72 τ	6.2	8.59	30.9	*	6.5 red	9.3	31.27
30 υ¹ dbl.	6.0	8.24	24.29	II. II, 80	neb.	8.43	19.31
32 υ²	5.9	8.22	24.33	II. II, 48	neb.	8.43	19.27
22 φ¹ dbl.	6.0	8.19	28.18	M. 67	cl.	8.45	12.15
23 φ² dbl.	6.2	8.20	27.20				

NOTES.

Præsepe— Is a beautiful cluster, visible to the naked eye; composed of stars from 6½ to the 10th magnitude; very easy with opera glass and instruments of small power; it is also called the " Bee Hive;" eight minutes east of it will be seen two small nebulæ. William Herschel, in 1784, saw another one there, but it was never seen afterward; it was perhaps a comet (Flam., Les Etoiles, page 333). (See Fig. 98.)

Fig. 98.—The Præsepe, or Bee Hive.

Scale—10″ = 1 inch.

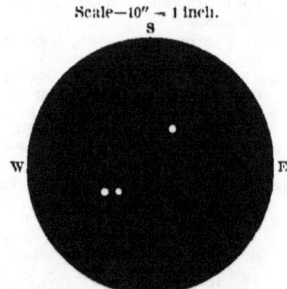

Fig. 99.—Trinary ζ in 1880.

ɩ (*Iota*) –Double; magnitudes, 4.5 and 7; distance, 30″; pale orange and blue; nice contrast.

ζ (*Zeta*) –Trinary; magnitudes 5.5-6.2 and 6.6; distance, 0″.8 and 5″.3; the first two revolve around each other in 60 years; the third revolves around the two others in 600 years; it is the most important system of the kind. (Fig. 99.)

φ² (*Phi*) —Double; magnitudes 6.0 and 6.5; distance, 4″.8; very brilliant pair; sometimes the two stars seem to be of equal magnitude; φ¹ (see φ) is also double; nice field.

θ (*Theta*) —Double; magnitudes 5.5 and 9; distance one minute exactly.

57 —Double; magnitudes 5.8 and 7; distance, 1″.4; close pair.

Σ 1298 —Double; magnitudes 6.5 and 9; distance, 4″.8; very nice pair; white and blue.

Σ 3121 —Binary; magnitudes 7.2 and 7.5; distance, 0″.7; revolution 39 years; very close pair; between β (*beta*) and 8.

M. 67 —Is a rich cluster, almost visible to the naked eye, composed of several hundreds of stars of the 10th and 11th magnitudes; see with opera glass near 60. There are several variables in this constellation; the principal ones are:

R.—Varies from 6.3 to 13th magnitude in 359 days (see north of β (*beta*).)

S.—Varies from 8th to 10½th magnitude in 9 days 11 hours 37 minutes (see between δ (*delta*) and γ (*gamma*).)

T.—Varies from 8.3 to 9.9 magnitude in 455 days (see about half way between δ (*delta*) and ξ (*zi*).)

Fig. 100.—Cluster Messier 67.

LEO.

Leo, or the Lion, is, according to Greek mythology, the famous lion killed by Hercules in the forest of Nemea, and placed among the constellations by Jupiter to commemorate the exploit of his son.

DESIGNA-TION.	MAGNI-TUDE.	POSITION		DESIGNA-TION.	MAGNI-TUDE.	POSITION	
		R. A. 1880 h. m.	DECL. ° ′			R. A. 1880 h. m.	DECL. ° ′
α dbl.	1.9	10. 2	+12.33	22 g	5.8	9.45	+24.50
β dbl.	2.1	11.43	15.15	6 h dbl.	5.7	9.26	10.16
γ bin.	2.2 yel.	10.13	20.27	52 k	6.0	10.40	14.49
δ trip.	2.8	11. 8	21.11	53 l	5.7	10.43	11.11
ε	3.0	9.39	24.20	51 m	6.0	10.40	19.31
ζ	3.3	10.10	24. 1	73 n	5.8	11.10	13.58
η	3.8	10. 1	17.21	95 o	6.0	11.50	+16.19
θ	3.4	11. 8	16. 5	p¹	5.9	10.48	— 1.29
ɩ bin.	4.0	11.18	11.12	61 p²	5.4	10.56	— 1.50
κ dbl.	4.8	9.18	26.42	62 p³	6.2	10.57	+ 0.39
λ	4.6 red	9.25	23.31	65 p⁴	5.8	11. 1	2.37
μ dbl.	4.2	9.46	26.34	69 p⁵	5.6	11. 8	0.36
ν	5.1	9.52	13. 1	49 dbl.	6.0	10.29	9.17
ξ	5.5	9.26	11.51	54 dbl.	4.5	10.49	25.25
o dbl.	3.9	9.35	10.26	71	7.4	11.12	18.32
π	5.2 red	9.54	8.37	72	5.0 org.	11. 9	23.45
ρ	4.0	10.26	9.55	75	6.0 org.	11.11	2.40
σ dbl.	4.2	11.15	6.41	92	5.8	11.35	22. 1
τ dbl.	5.2	11.22	+ 3.31	93 dbl.	4.5	11.42	20.54
υ dbl.	4.4	11.31	— 0.10	P. IX, 230	6.0	10.58	0.37
φ dbl.	4.3	11.11	— 3. 0	83 dbl.	7.0	11.21	3.40
χ	4.7	10.50	+ 7.59	88 dbl.	6.0	11.26	15. 3
ψ dbl.	5.5	9.37	14.35	90 trip.	6.0	11.28	17.28
ω bin.	5.9	9.22	9.36	R	6.V. org.	9.41	11.59
				•	6.5	11.29	11.35
31 A	5.0 org.	10. 2	10.36	M. 65	neb.	11.13	13.45
60 b	4.9 red	10.56	20.49	M. 66	neb.	11.14	13.38
59 c	5.0	10.55	6.45	H. III, 76	neb.	11.12	15.23
58 d	5.3	10.54	+ 4.16	H. I, 56-57	neb.	9.25	22. 2
87 e	5.2 red	11.24	— 2.21	H. I, 17-18	neb.	10.41	13.13
15 f	5.7	9.37	+30.32	M. 95	neb.	10.38	+12.19

NOTES.

α (Alpha) Regulus—Was observed and its position determined 2,120 years B. C., by the Babylonian astronomers; it was then in longitude 92° 30′; it is now in longitude 148° 9′, and it is by the change of longitude that Hipparchus discovered the precession of the equinoxes, 127 years B. C.; it was then in longitude 119° 50′.   Regulus is a double; magnitudes 1.9 and 8; distance, 2′ 57″; the companion is also double; distance, 3″. It has a motion of 27″ west in 100 years; Mr. Elkin, of Yale College, in 1887-88, obtained a parallax quite uncertain =0″.093±0″.048; if it was equal to only one-tenth of a second the distance between the two stars would be 163 billions 490 millions of miles, and if Regulus has the same attraction as our own sun, the companion would revolve around it in no less than 76,000 years. (Flam., Les Etolles, pages 354-5.)

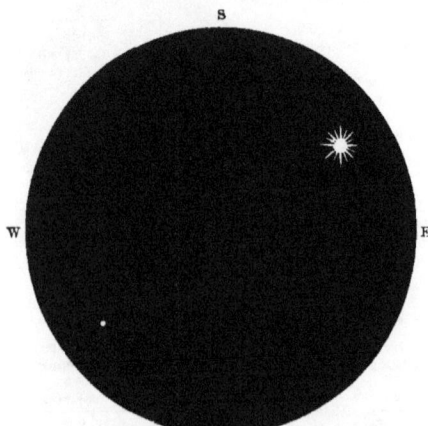

Fig. 101.—Double Star Regulus. Scale—100″=1 inch.   Fig. 102.—Binary γ

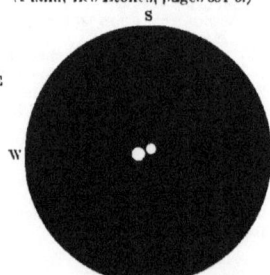

β (Beta) Denebola—Double; magnitudes 2.1 and 8; distance, 4′ 42″; nice field.

γ (Gamma) Algeiba—Binary; magnitudes 2.5 and 4; distance, in 1880, 3″.3; beautiful pair—one of the brightest; in the field will also be found two stars of the 7th and 9th magnitudes.

δ (Delta) Zosma—Is a triple star.

ζ (Zeta)—Double; magnitudes 3.3 and 6; distance, 5′ 19″; easy pair; an opera glass is sufficient.

ι (Iota)—Binary; magnitudes 4 and 7; distance, 2.″7; the main star is yellow, and the companion changes in color from greenish-yellow to dark blue; when it is of the darkest color it sometimes comes down as low as the 9th magnitude.

54—Double; magnitudes 4.5 and 7; distance, 6″.3; nice pair.

τ (Tau)—Double; magnitudes 5.2 and 7; distance, 94″; very easy pair.

88—Double; magnitudes 6 and 8; distance, 15″; system in rapid motion.

90—Triple; magnitudes 6, 7 and 8; distances, 3″.3 and 64″.

83—Double; magnitudes 7 and 8; distance, 30″; nice pair; white and pale rose.

ω (Omega)—Binary; magnitudes 5.9 and 7; distance, 0″.5 in 1880; white and blue; revolution 124 years.

R.—Is the principal variable of this constellation and varies from 5.8 to the 11th magnitude in 331 days; visible to the naked eye at its maximum.   It is between Regulus and *o (omicron)*; in the same field with 18 and 19, and its red color forms a striking contrast with these two stars, which are of the 6th and 7th magnitudes and perfectly white.

M. 65 and M. 66—Nebulæ 1′ 19″ apart from each other (see south of *θ (theta)* near 73).

Fig. 103.—Nebulæ M. 65, and M. 66.

II. I, 56 —Visible in the field south of θ (*theta*), is another double nebula 1½' wide and 3' long, south of λ (*lambda*).

II. I, 17—Is also a double nebula between θ (*theta*) and ρ (*rho*).

M. 95—Is also a double nebula; between θ (*theta*) and ρ (*rho*) is another nebula west of the above.

α (*Alpha*), η (*eta*), γ (*gamma*), ζ (*zeta*), μ (*mu*) and ε (*epsilon*), form what is called the "Sickle."

Fig. 104.—Nebula H. I, 56, in Lord Rosse's telescope.

Fig. 105.—Nebula M. 65, in Lord Rosse's telescope.

VIRGO. ♍

Virgo, or the Virgin, is a very old constellation, mentioned already by Eudoxus, Aratus, Hipparchus and Ptolemy, who called it "Parthenos" (the Virgin); it was also named Ceres, from the goddess of Agriculture; Themis, from the goddess of Justice; Astræa, from the daughter of Jupiter and Themis, goddess of Innocence and Purity, who was the last of the gods obliged to withdraw from the earth at the end of the "Golden Age," on account of the crimes of humanity; it was also sometimes named from Diana, of Ephesus; from Isis, of Egypt; from Minerva, mother of Bacchus, etc.   It is called Erigone by Virgil.

DESIGNA-TION.	MAGNI-TUDE.	POSITION		DESIGNA-TION.	MAGNI-TUDE.	POSITION	
		R. A. 1880 h. m.	DECL. ° '			R. A. 1880 h. m.	DECL. ° '
α (Spica) dbl.	1.5	13.19	—10.32	χ	5.2	12.33	— 7.20
β dbl.	3.5	11.44	+ 2.26	ψ	5.2	12.48	— 8.54
γ bin.	3.2	12.36	— 0.47	ω	6.0	11.32	+ 8.48
δ dbl.	3.4 red	12.50	+ 4. 3				
ε dbl.	2.8	12.56	+11.36	4 A¹	5.8	11.42	— 8.55
ζ dbl.	3.5	13.29	+ 0. 1	6 A²	6.1	11.49	+ 9. 7
η	3.9	12.14	0. 0	7 b	5.8	11.54	+ 4.19
θ trip.	4.6	13. 4	— 4.54	16 c	5.5	12.14	+ 3.58
ι	4.1	14.10	— 5.24	31 d¹	6.0	12.36	+ 7.28
κ	4.2 red	14. 6	— 9.43	32 d²	5.8	12.40	+ 8.20
λ	4.9	14.13	—12.49	59 e	5.5	13.11	+10. 3
μ	4.0	14.37	— 5. 8	25 f	6.0	12.31	— 5.10
ν	4.1	11.40	+ 7.12	g	6.0	13. 2	— 8.20
ξ	5.3	11.39	+ 8.55	73 h	5.8	13.27	— 9.32
o dbl.	4.2	11.59	+ 9.24	68 i	5.7 org.	13.29	—12. 5
π	4.8	11.55	+ 7.17	44 k, dbl.	6.0	12 53	— 3.10
ρ	5.0	12.36	+10.54	74 l	5.2 red	13.26	— 5.38
σ	5.3 red	13.42	+ 6. 6	82 m	5.8	13 37	— 8. 6
τ dbl.	4.4	13.56	+ 2. 8	n	6.8	13.49	— 8.58
102 ν¹	5.6	14.13	— 1.42	78 o	5.3	13.28	+ 4.16
103 ν²	6.8	14.16	— 1.26	90 p	5.6	13.49	— 0.55
φ dbl.	5.2	14.22	— 1.41	21 q	5.V.	12.28	— 8 48

VÍRGO—CONTINUED.

DESIGNATION.	MAGNITUDE.	POSITION R. A. 1880 h. m.	POSITION DECL. ° '	DESIGNATION.	MAGNITUDE.	POSITION R. A. 1880 h. m.	POSITION DECL. ° '
49	5.6	13. 2	—10. 6	*	7.0 red	11.52	+ 4.10
50	6.3	13. 3	— 9.41	*	7.5 red	14.19	+26.15
53 dbl.	5.3	13. 6	—15.33	*	8.0 org.	12.19	+ 1.27
61 dbl.	5.3	13.12	—17.39	*	7.0 org.	14.18	+ 8.38
63	5.6	13.17	—17. 6				
69	5.0	13.21	—15.21	R	7.V. org.	12.32	+ 7.39
70	5.5	13.23	+14.26	S	7.V. org.	13.27	— 6.35
75 dbl.	6.0	13.26	—14.45	T	8.V. red	12. 8	— 5.22
84 bin. (?)	5.5	13.37	+ 4. 9	U	8.V. org.	12.45	+ 6.12
86 dbl.	5.8	13.40	—11.49	V	8.V. org.	13.22	— 2.33
89	5.4	13.43	—17.32	X	7.V.	11.56	+ 9.44
96	6.9	14. 3	— 9.47	*	7.V.	12.32	+ 2.38
97	7.0	14. 7	+ 8.35	*	7.V.	13.57	— 1.48
109	4.5	14.40	+ 2.24	M. 60	neb	12.38	+12.13
4700 B. A. C.	5.5 red	14. 4	—15.44	II. IV, 8-9	neb. dbl.	12.31	+11.53
110	4.9	14.57	+ 2.34	M. 58	neb.	12.33	+12.29
P. XII, 142	4.V.	12.32	+ 2.31	M. 84	neb.	12.20	+13.33
P. XIII, 174	6.5	13.38	— 4.54	M. 86	neb.	12.21	+13.46
P. XIV, 12	5.0	14. 6	+ 2.58	M. 87	neb.	12.25	+13. 2
Lal. 23228	6.1	12.19	—10.56	M. 88	neb.	12.26	+15. 5
Lal. 25086	5.V.	13.28	—12.36	M. 90	neb.	12.31	+13.49
54 dbl.	6.3	13. 7	—17.56	M. 91	neb.	12.33	+14.26
84 dbl.	5.8	13.37	+ 4. 9	H. II, 74-75	neb. dbl.	12.47	+11.57
P. XII, 32 dbl.	6.0	12.12	— 3.16	M. 99	neb.	12.13	+15. 5
P. XII, 196 dbl.	6.5	12.46	— 9.38	H. I, 43	neb.	12.34	—10.57
P.XIII,127 dbl.	8.0	13.29	+ 0.21	H. I, 70	cl.	14.23	— 5.26
17 dbl.	7.0 red	12.16	+ 5.58	M. 61	neb. dbl.	12.16	+ 5. 8
*	6.5	12.32	+14.50				

NOTES.

β (Beta) is also called Zavijava, and ε (epsilon) Vindemiatrix.

γ (Gamma)—Binary; magnitude 3.0 and 3.2; distance, in 1880, 5"; one of the finest pairs; both yellow; we see its orbit from here nearly as it is in reality. We give below a table showing the position of the components and names of observers since 1718:

YEAR.	OBSERVERS.	DISTANCE.	ANGLE.
1718	Bradley and Cassini	6" ±	331°
1756	Tobie Mayer	6" ±	324°
1781	William Herschel	0"	311°
1803	William Herschel	4½"	300°
1820	John Herschel and South	3"	284°
1830	W. Struve and Dawes	1".8	262°
1836	Smyth, Dawes, Struve	0".4	140°
1840	Kaiser, Galle, Mädler	1".3	27°
1850	Wrottesley, Main, Jacob	2".8	356°
1860	Secchi, Knott, Dembowski	3".9	348°
1870	Duner, Wilson, Gledhill	4".5	342°
1880	Hall, Stone, Flammarion	5".0	337°

The perihelion arrived in 1836, and the two stars were so close to each other that they could not be separated; the revolution takes 175 years; the velocity of the companion is over 60 times more rapid at the perihelion than at the aphelion. This remarkable pair shows no signs of parallax, and the two stars must be at a great distance from each other; one is sometimes brighter than the other.

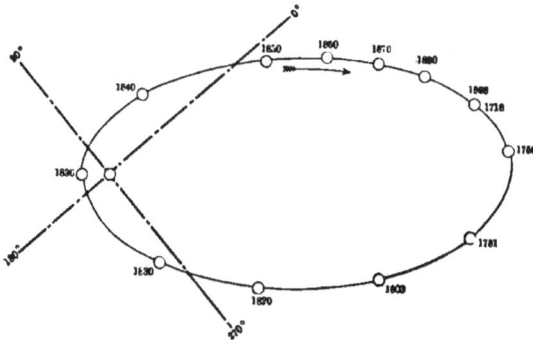

Fig. 106.—Orbit of γ

Σ 1757—Binary; magnitudes 8 and 9; distance, 2″; very close pair; white and yellow; revolution, 292 years.

θ (Theta)—Triple; magnitudes 4.5-9 and 10; distances, 7″ and 65″.

84—Double; magnitudes 5.8 and 8.5: distance, 3″.5; yellow and blue; suspected binary.

54—Double; magnitudes 6.3 and 7.5; distance, 5″.7; stationary for over 100 years.

17—Double; magnitudes 6.5 and 9; distance, 20″; rose and red; beautiful pair.

P. XII, 196—Double; magnitudes 6.5 and 9.5; distance, 33″; both orange.

P. XII, 32—Double; magnitudes 6.0 and 6.5; distance, 21″; nice pair.

P. XII, 127—Double; magnitudes 8 and 9; distance, 2″.3: delicate pair; very near ζ (zeta).

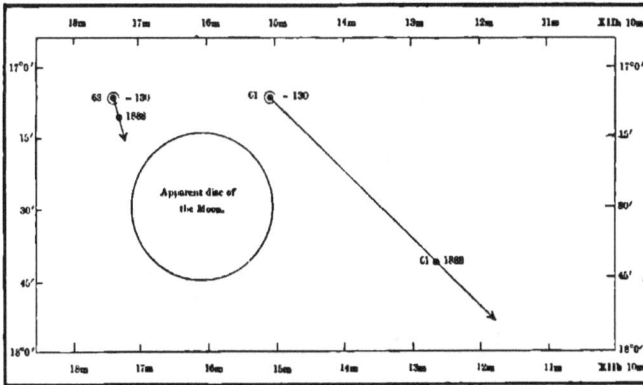

Fig. 107.—Proper motion of Star 61 since Hipparchus.

61—Is a star in rapid motion; the only one whose motion has been observed by the naked eye in comparing its position at different epochs with the surrounding stars; in Hipparchus' time it was near 63; since Hipparchus, 130 B. C. to 1888, it has moved 48′ S. W., or 1½ times the apparent diameter of the moon. (Fig. 107.)

Σ 1819—Binary; magnitudes 7 and 8; distance, 1″.2; very close pair; revolution, 380 years (see near 12).

This constellation contains more than 500 nebulæ. It is the richest of the heavens, containing more nebulæ than stars; we will note only the principal ones:

M. 60; M. 59; M. 58; H. II, 71; H. IV, 8; M. 84; M. 85; M. 86; M. 87; M. 88; M. 89; M. 90—North of ρ (rho); splendid field. (Fig. 108.)

Fig. 108.—A Field of Nebulæ in Constellation Virgo.

Fig. 109.—Nebulæ M. 60, H. 11, 71, and M. 59.

Fig. 110.—Nebulæ H. II, 75, and H. II, 74.

Fig. 111.—Double Nebula M. 61

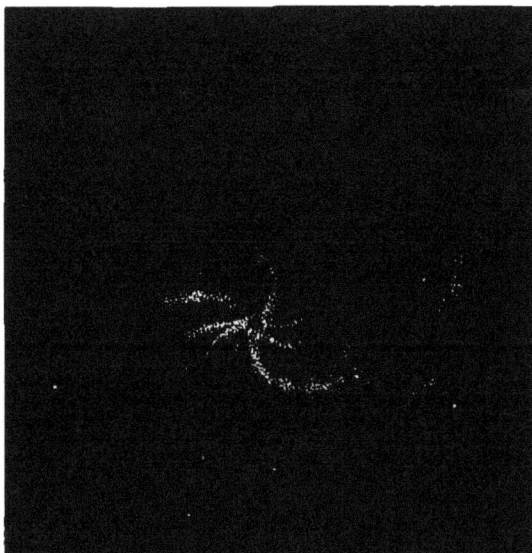

Fig. 112.—Nebula Messier 99, in Lord Rosse's telescope.

II. II, 75—Near ε (*epsilon*); oval nebula, comet-like, with three small stars in triangular shape. In the same field is another circular nebula with a double star near it of the 7th and 9th magnitudes. (Fig. 110.)

M. 61—Between 16 and 17 is a curious double nebula. (Fig. 111.)

M. 99—Near 6 of Coma Berenices is the splendid nebula, which, in Lord Rosse's telescope, looked like a lighted "pinwheel." (Fig. 112.)

II. I, 43—Between 196 and 23,675 of Corvus, is a nebula 4 minutes long by only 50 seconds wide.

II. I, 70—Between ι (*iota*) and μ (*mu*) is a nice cluster of blue stars, with a red star of the 8th magnitude near it.

LIBRA.

Libra, the Scales or the Balance, is generally considered as having been introduced during the reign of the Roman emperor Augustus Cæsar, to celebrate his justice; but it was already mentioned by Manetho, Egyptian historian, about 300 years B. C., and was taken out of the constellation Scorpio to form the twelfth sign of the zodiac.

DESIGNA-TION.	MAGNI-TUDE.	POSITION		DESIGNA-TION.	MAGNI-TUDE.	POSITION	
		R. A. h. m.	1880 DECL. ° '			R. A. h. m.	1880 DECL. ° '
α dbl.	3.0	14.44	−15.33	ξ²	5.7	14.50	−10.55
β dbl.	2.9 V.	15.11	8.56	o dbl.	6.4	15.14	15. 7
γ	4.4 yel.	15.29	11.24	6	5.5 red	14.41	27.25
δ	5.V.	14.55	8.02	11	5.4	14.45	1.48
ε	5.5	15.18	9.53	16	4.8	14.51	3.51
ζ quad.	5.8	15.21	16.18	49	5.6	15.54	16.11
η	5.9	15.37	15.17	37	5.5	15.28	9.39
θ	4.8 org.	15.47	16.22	28344 Lal.	5.6	15.28	8.47
ι dbl.	5.0	15. 5	19.29	48	5.4	15.51	13.56
κ	5.5	15.35	19.17	P.XIV,212trip.	6.3	14.50	20.52
λ	5.5	15.46	19.48	S	7.V.	15.15	19.57
μ dbl.	5.7	14.43	13.39	"	7.V. org.	15.37	10.52
ν dbl.	5.5 red	15. 0	15.47	Σ 1962 dbl.	6.0	15.32	8.24
ξ¹	6.1	14.48	11.24				

NOTES.

α (Alpha) *Kiffa Australis or Zuben el Genubi—Double;* magnitudes 3 and 6; distance, 3' 49″; easy pair with an opera glass.

β (Beta) *Kiffa Borealis or Zuben el Chameli—*Has a greenish color; very rare.

γ (Gamma) *Zuben Hakraki—*Was seen of the 6th magnitude by Hevelius; of 3d by Bayer; it is now of the 4½th magnitude.

ζ (Zeta)—Has in the same field three little stars of the 6th magnitude.

ι (Iota)—Triple; magnitudes 5.0–9 and 10; distances, 57″ and 1″.9; nice field.

P. XIV, 212—Double: magnitudes 6.3 and 7; distance, 15″; it is in rapid proper motion, 202″ in 100 years; direction S. E.; the smaller one does not go quite so fast as the other, and perhaps forms only a pair in perspective (Flam., Les Etoiles, page 388).

δ (Delta)—Is a variable of the shortest period, varying from 4.9 to 6.1 magnitude in 2 days 7 hours 51 minutes and 19 seconds.

There are three other variables in this constellation requiring good power to be followed.

SCORPIO. ♏

Scorpio, or the Scorpion. This constellation is very old, being mentioned already by Eudoxus, and its name comes from its shape and the brilliancy of its stars, which offer some resemblance to this venomous animal. In ancient times this constellation was extended as far as the Virgo. The "claws" were taken out to form the constellation Libra.

It is also supposed to be the scorpion which bit Orion at the time when he was near catching Diana, whom he was pursuing.

DESIGNA-TION.	MAGNI-TUDE.	POSITION R. A. 1880 h. m.	DECL. ° ′	DESIGNA-TION.	MAGNI-TUDE.	POSITION R. A. 1880 h. m.	DECL. ° ′
α (Antares) bin.	1.6 red	16.22	−26.10	2 A	5.2	15.47	−24.57
β trip.	2.5	15.58	19.29	2 b	5.3	15.44	25.23
γ	3.5 red	14.57	24.49	13 c¹	5.3	16. 5	27.37
δ	2.4	15.53	22.17	P. XVI, 31 c²	5.5	16. 5	28.19
ε	2.3	16.42	34. 4	19	5.1	16.13	23.53
ζ¹	5.8	16.45	42. 9	22	5.3	16.23	24.51
ζ²	3.6	16.46	42. 9	24	5.5	16.34	17.31
η	3.6	17. 3	43. 4	P. XV, 116	3.9	15.29	27.45
θ	2.1	17.29	42.56	P. XVI, 35	0.0	16.11	30.37
ι dbl.	3.3	17.39	40. 5	P. XVI, 55	5.8	16.15	38.55
κ	2.6	17.34	38.58	P. XVI, 92	5.7	16.22	34.27
λ	2.0	17.25	37. 1	P. XVI, 111	4.4	16.27	35. 1
μ¹	3.6	16.44	37.51	P. XVI, 236 dbl.	6.3	16.50	19.21
μ²	3.9	16.44	37.49	P. XVI, 255	5.7	16.53	31.58
ν dbl.	4.3	16. 5	19. 9	P. XVII, 137	4.5	17.27	38.33
ξ trin.	4.6	15.58	11. 3	P. XVII, 229	3.4	17.40	37. 0
ο	3.8	15.31	29.23	Σ 1999 dbl.	7.4	16. 1	11. 7
π	3.4	15.52	25.46	T (1860)	7.V.	16.10	22.41
ρ	4.5	15.49	28.52	*	6.V.	16.49	32.58
σ dbl.	5.4	16.14	25.18	*	8.0 blood red	16.32	32. 8
τ	3.2	16.28	27.58	*	8.0 red	17.32	41.33
υ	3.2	17.23	37.12	*	6.0 red	16.28	35. 0
χ¹	5.6	16. 7	11.32	M. 80	cl.	16.10	22.41
ψ	5.2	16. 5	9.45				
ω¹ dbl.	4.4	16. 0	20.20				
ω²	4.6 red	16. 1	20.33				

NOTES.

α (Alpha) *Antares—*Is one of the finest double stars; magnitudes 1.6 and 7; distance, 3″.3; red and emerald; it requires a good telescope to separate them; it is most likely a binary.

β (Beta)—Triple; magnitudes 2.5–10 and 5.5; distances, 0″.9 and 13″; the first two form a difficult pair; the close companion was discovered at Chicago by Mr. Burnham.

Scale—25″=1 inch.

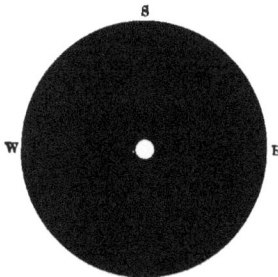

Fig. 113.—Double Star Antares.

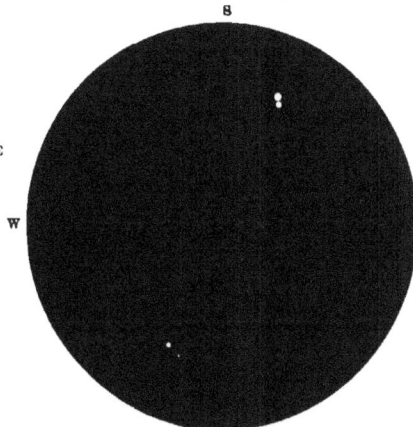

Fig. 114.—Quadruple Star ν

ν (*Nu*)—Quadruple; composed of two double stars of the 4th and 7th magnitudes, at 40″ distance; very easy pair; the star of the 7th magnitude was noted double in 1846 by Mr. Mitchell, at Cincinnati, and the brightest by Mr. Burnham, at Chicago, in 1874; the two doubles are very close and difficult pairs; the first is composed of two stars of the 4th and 5th magnitudes; distance only 1″; the second, of two stars of the 7th and 8th magnitudes; distance 1″.9. (Fig. 114.)

σ (*Sigma*)—Double; magnitudes 3.4 and 9; distance, 20″.

ω (*Omega*)—Double; magnitudes 4.5 and 4.5; distance, 14½′; they are far enough apart to be separated with the naked eye; very interesting pair.

μ (*Mu*)—Double; magnitudes 3.6 and 3.9; distance, 8′; easy with an opera glass.

ξ (*Zi*)—Trinary; magnitudes 5.0-5.2 and 7.5; distances, 1″.3 and 7″.3, in 1880; the first two revolve around each other in 96 years. (Fig. 115.)

P. XVI, 35—Double; magnitudes 6 and 8; distance, 23″; nice pair.

Σ 1999—Double; magnitudes 7.4 and 8.1; distance, 10″.

Anonyma—Between ε (*epsilon*) and τ (*tau*) is a very red star, called "the drop of blood" by Sir John Herschel; it is of the 8th magnitude.

M. 80—Is a cluster composed of a large number of small stars, and in the same field are the three variables, R., S. and T. See between Antares and β (*beta*).

Scale—25″=1 inch.

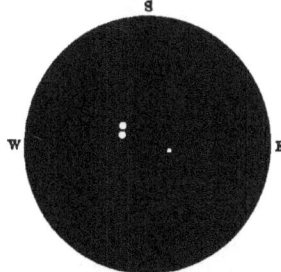

Fig. 115.—Trinary ξ

T (1860)—T, in the center of the above nebula, is generally of the 10th magnitude; but the 21st of May, 1860, it was seen of the 6½th magnitude, nearly visible to the naked eye; and the 10th of June it came down to its ordinary magnitude.

SAGITTARIUS.

Sagittarius, or the Archer, represents, according to the Greeks, the Centaur Cheiron, to whom they attribute the invention of the celestial sphere. He was the preceptor of Peleus, Achilles and Diomed. Wounded by a poisonous arrow, and suffering terribly, he asked to be put to death; but as he was immortal, the gods placed him among the stars.

This constellation is said to have been designed by Cleostratus of Tenedos, in the 6th century B. C. It is also mentioned by Eudoxus.

DESIGNA-TION.	MAGNI-TUDE.	POSITION R. A. 1880 h. m.	DECL. ° '	DESIGNA-TION.	MAGNI-TUDE.	POSITION R. A. 1880 h. m.	DECL. ° '
α	4.0	19.15	—40.51	43 d	5.6	19.11	19.10
β¹ dbl.	3.8	19.14	44.41	54 e¹ trip.	5.5	19.34	16.34
β²	4.4	19.14	45. 1	55 c²	5.4	19.36	16.24
γ	2.8 yel.	17.58	30.25	56 f	5.2	19.39	20. 3
δ	2.8 red	18.13	29.53	61 g	5.3	19.51	15.48
ε	2.2	18.16	34.26	51 h¹	5.V.	19.29	24.59
ζ dbl.	3.1	18.55	30. 3	52 h²	4.7	19.29	25. 9
η	3.3 yel.	18. 9	36.48	3 X	4.V.	17.40	27.47
θ¹	4.5	19.52	35.36	W	5.V.	17.57	29.35
ι	4.3	19.47	42.11	4 dbl.	5.4	17.52	23.48
κ¹	5.5	20.14	42.26	9	6.0	17.56	24.22
λ	2.7 yel.	18.21	25.29	21	5.1 red	18.18	20.36
13 μ¹ dbl.	4.3 org.	18. 7	21. 5	29	5.5	18.43	20.28
15 μ²	5.8	18. 8	20.46	P. XVII, 294	5.4	17.50	30.14
ν¹ dbl.	5.0 red	18.47	22.54	P. XVII, 359	5.1	17.59	28.28
ν²	5.1 red	18.48	22.49	P. XVII, 367	5.9	18. 1	30.45
ξ²	3.5 red	18.51	21.16	P. XVIII, 24	5.1	18.10	27. 6
ο	3.8 yel.	18.57	21.55	P. XVIII, 146	5.2	18.35	35.46
π	3.1	19. 3	21.13	8310 Lac.	5.0	19.56	38.16
ρ¹	4.2	19.15	18. 4	R	7.V. org.	19.10	19.31
σ dbl.	2.4 V.	18.48	26.27	T	8.V. org.	19. 9	17.11
τ	3.6 yel.	18.59	27.51	U	7.V. org.	18.25	19.13
υ	4.9	19.15	16.11	*	8.0 red	20.21	28.39
φ	3.7	18.38	27. 7	*	6.5 red	19.27	16.38
47 χ¹	5.4	19.18	24.44	*	7.0 red	20. 0	27.34
49 χ²	5.6	19.18	24.12	Temp. of 1690	....	19. ±	20. ±
ψ	5.4	19. 8	25.28	M. 22	cl.	18.29	24. 0
ω	5.1	19.48	26.37	M. 25	cl.	18.25	19. 9
				M. 8	cl.	17.57	24.22
60 A	5.3	19.52	26.31	M. 20	neb.	17.55	23. 2
59 b	4.6	19.50	27.29	M. 21	cl.	17.57	22.31
62 c	4.7 yel.	19.55	28. 3	II. VII, 30	cl.	18. 6	21.36

NOTES.

β¹ (*Beta¹*) and β² (*Beta²*)—Magnitudes 3.8 and 4.5; distance, 22'. β¹ (*beta¹*) is itself double; its
  companion, at 29'' distance, was noted of the 9th magnitude by Piazzi; of 8th magnitude
  by John Herschel, and of 6¾th magnitude by Gould.

ε (*Epsilon*)—Is also called Kaus Australis.

ξ¹ (*Zi¹*) and ξ² (*Zi²*)—Magnitudes 3½ and 5; distance, 29'.   }
ρ¹ (*Rho¹*) and ρ² (*Rho²*)—Magnitudes 4¼ and 6; distance, 28'.   } All visible
χ¹ (*Chi¹*) and χ² (*Chi²*)—Magnitudes 5.5 and 5.6; distance, 31'.   } to the
θ¹ (*Theta¹*) and θ² (*Theta²*)—Magnitudes 4½ and 5½; distance, 35'.   } naked eye.

51 and 52—Magnitudes 4.7 and 5; distance, 14'.   51 varies from 4.5 to 6.7 magnitude.

ν (*Nu*)—Double; magnitudes 5.0 and 5.1; distance, 12'.   This pair was already visible with
  the naked eye 2,000 years ago; Ptolemy took it for a double nebula; it was not well
  defined then.

μ¹ (*Mu¹*)—Triple; magnitudes 4.3-9 and 10; distances, 40'' and 45''; in large telescopes another
  star of 13th magnitude at 15'' is also seen, which really makes μ¹ (*mu¹*) a quadruple star.

54—Double; magnitudes 5½ and 8; distance, 28''; nice field of small stars.

21—Double; magnitudes 5.1 and 9; distance, 2''; very close pair; orange and blue.

σ (*Sigma*)—Double; magnitudes 2.5 and 9; distance, 5'; the main star is also variable.

This constellation, traversed by the "Milky Way," is rich in nebulæ; the principal are:

M. 8—A beautiful cluster, visible to the naked eye; in a large field could be seen a triple star
  followed by an agglomeration of stars offering two *foci* of condensation; splendid sight.

M. 21—Is another cluster, larger but not so bright; near the center is a double star of the
  9th magnitude.

X.—Varies from the 4th to the 6th magnitude in 7 days 17 minutes and 42 seconds.

W.—Varies from the 5th to the 6½th magnitude in 7 days 14 hours 15 minutes and 34 seconds.

U.—Varies from the 7th to the 8th magnitude in 6 days 17 hours 53 minutes and 1 second.

In this constellation, between γ (*gamma*) and 3 X, there is a curious black hole in the
  "Milky Way." (See Fig. 60, page 28.)

West of μ (mu), Mr. Pickering, of Cambridge, noticed, on the 28th of August, 1880, a star of the 8th magnitude, offering a spectrum similar to the one of the temporary of Corona Borealis, indicating an incandescent atmosphere; perhaps it is a bright nebula already condensed (Flam., Les Etoiles, page 415).

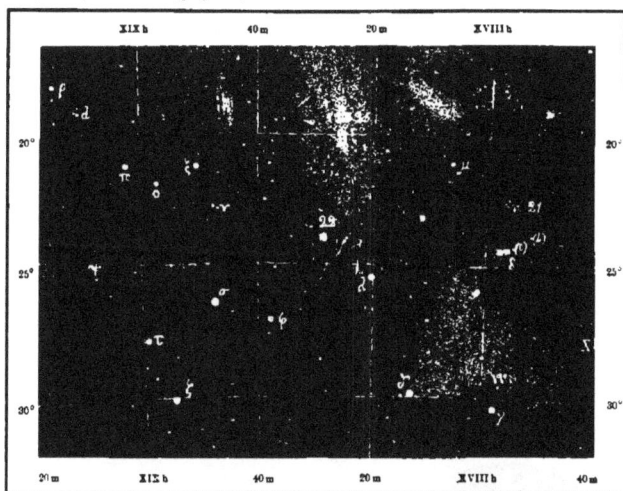

Fig. 116.—A Field of Nebulæ in the Constellation Sagittarius.

T (1690)—The Jesuit astronomers of the observatory of Pekin noted, the 28th of September, 1690, a star of the 4th magnitude; the 4th of October it was hardly visible, and it soon disappeared. This observation was found by Mr. Schiaparelli only some years ago; the position was not well defined (Flam., Les Etoiles, page 414).

CAPRICORNUS.

Capricornus, the Goat, also called the Sea-Goat, was the god Mentes of the Egyptians, and the object of several stories in Greek mythology. It is mentioned by Eudoxus, and is one of the oldest constellations.

DESIGNA-TION.	MAGNI-TUDE.	POSITION		DESIGNA-TION.	MAGNI-TUDE.	POSITION	
		R. A. h. m.	1880 DECL. ° '			R. A. h. m.	1880 DECL. ° '
α¹	4.5 yel.	20.11	—12.53	τ	5.6	20.33	—15 22
α² trip.	3.6 yel.	20.11	12.55	υ	5.7	20.33	18.34
β dbl.	3.2	20.14	15.10	φ	5.5	21. 9	21. 9
γ	3.7	21.33	17.12	χ trip.	5.4	21. 2	21.41
δ	2.8	21.40	16.40	ψ	4.3	20.39	25.42
ε dbl.	4.7	21.30	20. 0	ω	4.1	20.45	27.22
ζ	3.7	21.20	22.56	24 A	4.8	21. 0	25.30
η	5.1	20.58	20.20	36 b	4.7	21.22	22.20
θ	4.1	20 59	17.43	46 c¹	5.5	21.39	9.38
ι	4.4	21.16	17.21	47 c²	6.4	21.40	9.50
κ	5.0	21.36	19.25	29	5.7	21. 9	15.40
λ	5.7	21.40	11.56	30	5.5	21.11	18.29
μ	5.4	21.47	14. 7	33	5.7	21.17	21.22
ν	5.2	20.14	13. 8	41	5.8	21 35	23.48
ξ	6.3	20. 6	12.58	42	5.6	21.35	14.35
υ dbl.	6.3	20.23	18.59	S	7.V.	20.10	21.42
π dbl.	5.5	20.20	18.37	*	7.0 red	20.10	21.41
ρ dbl.	5.3	20.22	18.13	M. 30	cl.	21.34	23.43
σ dbl.	5.6 yel.	20.12	19.30	M. 72	cl.	20.47	12.59

NOTES.

Two thousand years ago the sun was in this constellation the 21st of December and in the constellation of Cancer the 21st of June; since that time the geographers indicate a line 23° 28′ north, and another 23° 28′ south of the equator, and call the first Tropic of Cancer and the other Tropic of Capricorn; by the effects of the precession of the equinoxes it is now Sagittarius, which is the most southern constellation of the Zodiac, and the Gemini, the northern one; still the old names keep going on.

α¹ (*Alpha*¹) and α² (*Alpha*²)—Magnitudes 3.6 and 4.5; distance, 6′ 16″ in 1880; these stars are going away from each other at the rate of 7″ for 100 years; in Hipparchus' time they were only 4 minutes apart; our Fig. 118 shows the separation of the two stars for 3,000 years.

α² (*Alpha*²) *Secunda Giedi*—Has a close double companion, in large telescopes; distances, 6″ and 1″.5.

β² (*Beta*²)—Double; magnitudes 3.2 and 7; distance, 3′ 25″; easy pair, with a star of the 8th to 9th magnitude in the field forming a nice triangle.

β¹ (*Beta*¹) is a double; magnitudes 7 and 9; distance, 0″.85; discovered by Mr. Barnard in 1883.

46 c¹—Double: magnitudes 5.5 and 7; distance, 3′; visible with an opera glass.

ρ (*Rho*)—Double; magnitudes 5.3 and 7.5; distance, 4′; the main star is itself a close pair, having a companion of the 9th magnitude at 3″.8 only.

σ (*Sigma*)—Double; magnitudes 5.6 and 10; distance, 54″; orange and lilac; easy pair.

ο (*Omicron*)—Double; magnitudes 6.3 and 7; distance, 22″; nice pair; both bluish.

π (*Pi*)—Double; magnitudes 5.5 and 8; distance, 3″.4; delicate pair.

M. 30 Is a nice cluster east of ζ (*zeta*); a star of the 6th magnitude appears in the field.

Fig. 117.—Cluster M. 30.

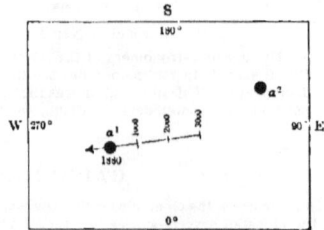

Fig. 118.—Separation of α¹ and α².

M. 72—Between ν (*nu*) of Aquarius and τ (*tau*) is also a nebula nearly 2 minutes in diameter, composed entirely of small stars defined for the first time by Herschel; a star of the 6th magnitude appears at 30′ from it.

This constellation contains some variables, too small for common telescopes.

R.—Varies from the 9th to 14th magnitude in about 1 year.

S.—Varies from the 7th to 8½th; time not yet known.

T.—Varies from the 9th to 14th in 274 days.

U.—Varies from the 10½th to 14th in 450 days.

AQUARIUS.

Aquarius, the Water Bearer or Waterman, is, according to Greek mythology, Ganymede, the beautiful Phrygian boy, son of Tros, carried by the eagle of Jupiter to heaven, where he took the place of Hebe as cupbearer of the gods.

It is one of the oldest constellations, already mentioned by Eudoxus.

| DESIGNATION. | MAGNITUDE. | POSITION | | DESIGNATION. | MAGNITUDE. | POSITION | |
		R. A. 1880 h. m.	DECL. ° '			R. A. 1880 h. m.	DECL. ° '
a dbl.	2.7	22. 0	-- 0.54	101 b³	4.5	23.27	−21.34
β dbl.	2.6	21.25	− 6. 6	86 c¹	4.4	23. 0	−24.23
γ dbl.	3.9	22.15	− 1.59	88 c²	3.7	23. 3	−21.49
δ	3.2	22.48	−16.28	89 c³	4.9	23. 3	−23. 6
ε	3.8	20.41	− 9.56	25 d	5.5	21.33	∔ 1.42
ζ bin.	3.5	22.23	− 0.38	38 e	5.6	22. 4	−12.01
η	4.1	22.29	− 0.44	53 f, dbl.	5.8	22.20	−17.21
θ	4.3	22.11	− 8.23	66 g¹	4.9	22.37	−19.28
ι	4.4	22. 0	−14.27	68 g²	5.4 yel.	22.41	−20.14
κ	5.2	22.32	− 4.50	83 h, dbl.	5.4	22.50	− 8.20
λ	3.6 red	22.46	− 8.13	106 i¹	5.2	23.38	−18.56
μ	5.0	20.46	− 9.26	107 i², dbl.	5.4	23.40	−19.21
ν	4.7	21.03	−11.52	108 i³	5.1	23.45	−19.34
ξ	5.0	21.31	− 8.24	1 trip.	5.6	20.33	∔ 0. 4
ο	4.9	21.57	− 2.44	3	4.8	20.41	− 5.28
π	4.9	21.19	+ 0.46	5	5.8	20.46	− 5.57
ρ	5.6	22.14	− 8.25	7	5.9	20.50	−10. 9
σ	5.1	22.24	−11.17	12 dbl.	5.7	20.58	− 6.18
69 τ¹, dbl.	5.8	22.41	−14.41	29 dbl.	6.0	21.56	−17.32
71 τ⁴, dbl.	4.2 red	22.43	−14.14	41 dbl.	5.8	22. 8	−21.40
υ	5.7	22.28	−21.19	46090 Lal.	6.V.	23.26	−11.49
φ	4.1 org.	23. 8	− 6.41	94 dbl.	5.5 yel.	23.13	−14. 7
χ	5.3 red	23.11	− 8.23	97	5.3	23.16	−15.42
ψ¹ dbl.	4.1 yel.	23.10	− 9.44	P. XXII, 250	5.9	22.49	− 5.38
ψ²	4.2	23.12	− 9.50	Σ 2809 dbl.	6.0	21.31	− 0.53
ψ³	4.8	23.13	−10.16	R	7.V. red	23.38	−15.57
ω¹	5.2	23.34	−14.53	S	8.V. org.	22.51	−20.59
ω² dbl.	4.7	23.36	−15.12	T	7.V. org.	20.44	− 5.35
				*	6.5 red	21.40	− 2.46
103 A¹	5.8	23.35	−18.41	*	7.0 red	22.53	−25.48
104 A²	5.0	23.36	−18.29	M. 2	cl.	21.27	− 1.21
98 b¹	3.9	23.17	−20.45	H. IV, 1	neb.	20.58	−11.50
99 b²	4.4	23.20	−21.18				

NOTES.

The principal stars are called: a (*alpha*) Sadalmelik; β (*beta*) Sadalsund; and δ (*delta*) Skat.

Near δ (*delta*) Tobie Mayer observed Uranus the 26th of September, 1756, in this constellation; had he known it was a planet he would have had the honor of discovering it twenty-five years before William Herschel.

ζ (*Zeta*)—Binary; magnitudes 3.5 and 4.4; distance, 3".5; nice pair; revolution 800 years—perhaps over 1,000 years.

83 h—Double; magnitudes 5.4 and 7.5; distance, 4'; easy pair.

ψ¹ (*Psi* ¹)—Double; magnitudes 4.1 and 9; distance, 50"; yellow and blue; easy pair; three stars in the field. The small star is a close double, discovered by Mr. Burnham; distance, 0".25.

τ¹ (*Tau*¹)—Double; magnitudes 5.8 and 9; distance, 28"; elegant pair.

94—Double; magnitudes 5.5 and 7.5; distance, 14"; rose and light blue; nice pair.

53 f—Double; magnitudes 5.8 and 6; distance, 8"; easy pair.

107 i¹—Double; magnitudes 5.5 and 7.5; distance, 5".6; white and purple.

Fig. 119.—Binary Star ζ

41—Double; magnitudes 5.8 and 8.5; distance, 4".8; topaz and blue; beautiful pair.

12—Double; magnitudes 5.7 and 8.5; distance, 2".8; close and delicate pair.

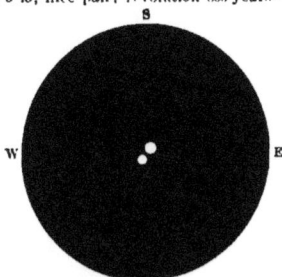

40090 Lal.—West of the three stars ψ (psi) is a variable from the 5½th to 8th magnitude; not visible to the naked eye since 1878; noted of 6th magnitude by Heis and Argelander; of 6½ by Lalande; period of variation not yet determined.

R.—Near ω (omega), varies from the 6th to the 11th magnitude in 388 days. There are other variables too small for common telescopes.

M. 2—Very fine "amas" discovered by Maraldi in 1746. When W. Herschel saw it through his 40-foot telescope he found there thousands and thousands of stars; diameter 3′ (see between β (beta) and 25). (Figs. 120 and 121.)

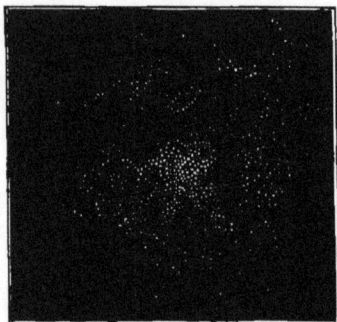

Fig. 120.—M. 2 in common telescope.

Fig. 121.—Cluster M. 2, in powerful telescopes.

Fig. 122.—Nebula H. IV, 1.

II. IV, 1—Is the beautiful nebula somewhat resembling Saturn in large telescopes; if this nebula was only as far from us as 61 of Cygnus this gaseous globe would be 264 billions times larger than the sun, or 338 quadrillions 896 trillions and 800 billions times larger than the earth; it measured 28″ in length and 18″ in width, and shines with a bluish color; is composed chiefly of nitrogen and hydrogen and is most likely a world in formation. In 1794 Lalande noted it as a star of 7½th magnitude, but W. Herschel, in 1782, already recognized it as a nebula (Revue d'Ast., 1882; page 291); it is between ν (nu) and ε (epsilon). (Fig. 122.)

THE SOUTHERN CONSTELLATIONS.

PLATE I.

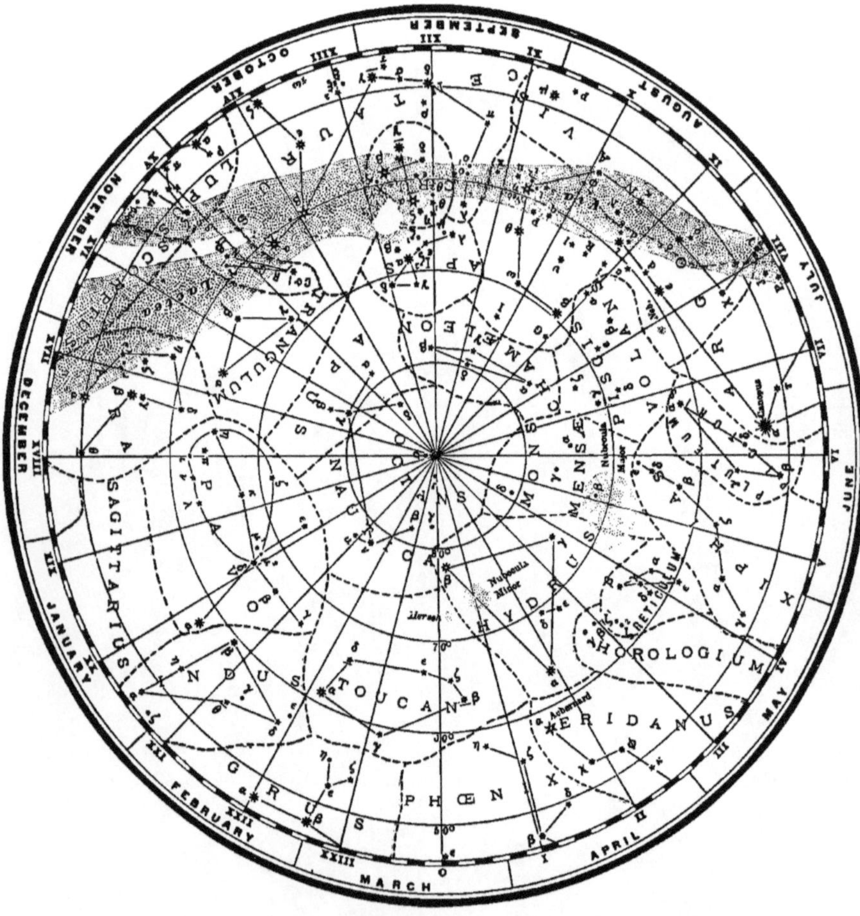

MAGNITUDE OF STARS.	✳ to ✳ 1st Magnitude.
	✴ to ✴ 2nd "
	✦ .... 3rd "
	✴ .... 4th "
	• .... 5th and under.
	✦ ✹ Clusters and Nebulæ.

Stars underlined are double or multiple.

CONSTELLATIONS SOUTH OF THE ZODIAC.

ORION.

This beautiful constellation, mentioned by Job (ix, 9), by Homer and Hesiod, is certainly one of the oldest of the heavens.

Orion was a mighty giant and hunter, the subject of many fables. He lost his sight in attempting to carry off the daughter of Œnopion, but he regained it by exposing his eyeballs to the rays of the rising sun; he was afterward loved and carried off by Aurora; this made the gods angry, and Diana killed him. According to others Diana herself was in love with him and killed him while he was swimming, mistaking his head for some other distant object on the water, which was pointed out to her by Apollo, who was indignant at his sister's love. According to another story, he was killed by a scorpion.

DESIGNA- TION.	MAGNI- TUDE.	POSITION			DESIGNA- TION.	MAGNI- TUDE.	POSITION		
		R. A. h. m.	1860	DECL. ° ′			R. A. h. m.	1860	DECL. ° ′
α dbl.	1.V. org.	5.49		+ 7.23	16 h	5.9	5.03		+ 9.40
β dbl.	1.3	5. 9		− 8.20	14 i, bin.	5.9	5. 1		+ 8.20
γ dbl.	2.0	5 19		+ 6.14	74 k	5.8	6.10		+12.18
δ	2.V.	5.26		− 0.23	75 l	6.0	6.10		+ 9.59
ε dbl.	2.0	5.30		− 1.17	23 m, dbl.	5.4	5.17		+ 3.25
ζ trip.	2.0	5.35		− 2. 0	33 n1, dbl.	6.0	5.25		+ 3.12
η bin.	3.5	5.18		− 2.30	38 n2,	5.8	5.28		+ 3.41
θ mult.	4.8	5.29		− 5.28	22 o, dbl.	5.1	5.16		− 0.30
ι trip.	3.0	5.30		− 5.59	27 p	5.6	5.18		− 1.01
κ	2.8	5.42		− 9.43	11	5.0	4.58		+15.14
λ trip.	3.5	5.29		+ 9.51	15	5.3	5. 3		+15.27
μ	4.7	5.56		+ 9.39	31 dbl.	5.1 org.	5.24		− 1.11
ν	4.7	6. 1		+14.47	52 dbl.	5.7	5.42		+ 6.26
ξ	4.8	6. 5		+14.14	56	5.8	5.46		+ 1.49
o1	5.7 org.	4.46		+14. 2	60	5.7	5.53		+ 0.32
o2	5.0	4.50		+13.19	5	5.V. org.	4.47		+ 2.18
π1	5.0	4.48		+ 9.58	9419 Lal.	6.2	4.54		+ 3.26
π2	4.7	4.44		+ 8.42	9581 Lal.	6.V.	4.50		+ 1. 1
π3	3.1	4.43		+ 6.45	10492 Lal.	6.V.	5.28		+10.10
π4	3.7	4.45		+ 5.24	10527 Lal., dbl.	5.3	5.29		− 6.05
π5	3.7	4.48		+ 2.15	11382 Lal.	5.2	5.54		− 3. 5
π6	4.7	4.52		+ 1.32	12104 Lal.	5.2	6.14		− 2.54
ρ dbl.	5.1	5. 7		+ 2.43	Σ 700, dbl.	8.0	5.17		+ 0.59
σ trip.	4.2	5.33		− 2.40	Σ 743, dbl.	7.0	5.29		− 4.30
τ trip.	4.4	5.12		− 6.58	Σ 750, dbl.	6.0	5.30		− 4.27
υ	5.1	5.26		− 7.24					
37 φ1	5.0	5.28		+ 9.24	R	8.V. org.	4.52		+ 7.57
40 φ2	4.5 yel.	5.30		+ 9.14	S	8.V. red	5.23		− 4.47
54 χ1	4.7	5.47		+20.16	*	8.0 very red	5. 4		+ 5.40
62 χ2	5.0	5.57		+20. 8	*	6.5 red	4.59		+ 1. 1
25 ψ1	5.4	5.19		+ 1.44	*	7.3 red	6. 5		+21.54
30 ψ2, dbl.	5.0	5.21		+ 2.59	*	6.5 org.	4.40		+ 7.35
ω	5.0	5.33		+ 4. 3	*	6.7 org.	5. 4		− 0.43
					*	6.5 org.	5.30		+10.58
32 A, bin. (?)	4.8	5.24		+ 5.51	*	7.0 org.	5.56		− 5. 8
51 b	5.5 org.	5.36		+ 1.25	*	6.5 org.	6.13		+14.42
32 c	5.2	5.29		− 4.55	*	6.5 org.	6.19		+14.47
49 d	5.2	5.33		− 7.17	M. 42	neb.	5.29		− 5.28
29 e	4.4	5.18		− 7.55	H. VII, 4	cl.	5. 4		+16.33
69 f1	5.7	6. 5		+16. 9	H. V, 23	neb.	5.36		− 1.55
72 f2	5.7	6. 8		+16.10	M. 78	neb.	5.41		+ 0. 1
6 g	6.0	4.47		+11.13	H. VIII, 24	cl.	6. 2		+13 58

NOTES.

This is the finest and richest constellation of the heavens, containing two stars of the 1st magnitude, four of the 2d magnitude, seven of the 3d magnitude and twelve of the 4th magnitude.

The same illusion which we have noticed in the Pleiades applies to the three stars λ (*lambda*), φ¹ (*phi¹*) and φ² (*phi²*); in looking at this triangle nobody would think that the moon could be inserted in it; but, as the distance from λ (*lambda*) to φ¹ (*phi¹*) is 27′, and the distance from φ¹ (*phi¹*) to φ² (*phi²*) 33′, it is a positive fact.

The principal stars of this constellation are named: α (*alpha*) Betelgeuse, β (*beta*) Rigel, γ (*gamma*) Bellatrix, δ (*delta*) Mintaka, and ε (*epsilon*) Alnilam. At the time of Bayer, Betelgeuse was brighter than Rigel, now it is the contrary.

α (*Alpha*)—Double; magnitudes 1.5 and 9; distance, 2′ 40″; yellow and blue. Several other nearer stars with the Lick telescope. Mr. Elkin tried the parallax of this nice star, in 1887-88, and found a negative one (−0″.009±0″.047).

β (*Beta*)—Double; magnitudes 1.3 and 9; distance, 9″.5; difficult pair on account of the brilliancy of Rigel; the companion is blue, the main star is very white; the best time to separate them in common telescopes is by moonlight or at twilight. Rigel has another companion of the 14th magnitude, discovered by Mitchell, at Cincinnati, in 1846, at 44″ distance; Mr. Burnham found the nearest companion double at the remarkably short distance of two-tenths of a second, in 1878, at Chicago.

(χ¹ *Chi¹*)—Double; magnitudes 4.7 and 6; distance, 32′; visible to the naked eye.

χ² (*Chi²*)—Double; magnitudes 5.0 and 6; distance, 28′; visible to the naked eye.

22—Double; magnitudes 5.0 and 6.0; distance, 4′; easy pair with an opera glass.

θ¹ (*Theta¹*) θ² (*Theta²*)—Magnitudes 5.0 and 5.5; distance, 2′ 15″; are visible with an opera glass.

θ¹ (*Theta¹*)—Is quadruple; magnitudes 5-6-7 and 8; distances from 9″ to 21″; splendid in common telescopes; multiple in large telescopes. It is also known as the *Trapezium of Orion*. Fig. 124 is taken from Mr. Burnham's diagram in *monthly notices of the R. A. S., Vol. XLIX., No. 6.* Mr. Barnard discovered a faint star, indicated by a small cross, which Mr. Burnham has not satisfactorily seen.

θ² (*Theta²*)—Double; magnitudes 5.5 and 6.5; distance, 52″; nice pair.

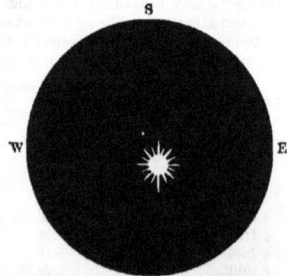

Fig. 123.—Double Star Rigel β

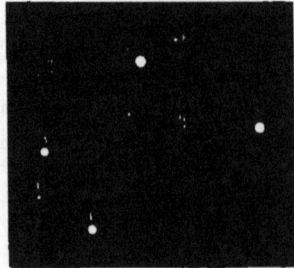

Fig. 124.—Trapezium of Orion.
Scale, 1 inch · 16″.

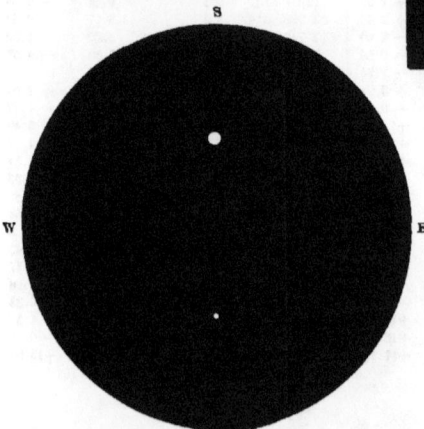

Fig. 125.—Double Star δ

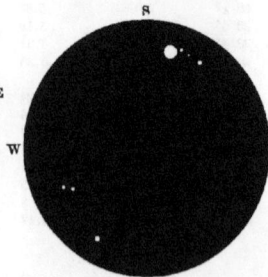

Fig 126.—Double Triple Star σ
Scale, 200″ −1 inch.

ι (*Iota*)—Triple; magnitudes 3.0-8½ and 11; distances, 11″ and 49″; 6′ south from ι (*iota*) there is an easy pair; magnitude 5.8 and 6.3; distance, 36″; the components seem to be variable; Lalande noted the first one of the 7th magnitude and the other of the 8th magnitude; Struve noted them 5.6 and 6.5.

δ (*Delta*)—Double; magnitudes 2.6 and 7; distance, 53″; very nice pair.

42—Double; magnitudes 5.6 and 6; distance, 5″; beautiful field with small power.

23—Double; magnitudes 5.4 and 7; distance, 32″; white and blue; nice pair.

σ (*Sigma*)—Triple; magnitudes 4.2-8 and 7; distances, 12″ and 42″; it is a double triple; three other small stars of the 8th and 9th magnitude appearing in the field; beautiful sight. Mr. Burnham discovered in 1888 the main star to be a very close pair; distance only 0″.25.

λ (*Lambda*)—Double; magnitudes 3.5 and 6; distance, 4″.5; another small companion appears in the field.

ρ (*Rho*)—Double; magnitudes 5.1 and 9; distance, 6″.8; orange and blue; nice pair.

ζ (*Zeta*)—Double; magnitudes 2.0 and 6.5; distance, 2″.5; difficult pair; the companion is quite dark; Struve called it "Olivaceasubrubiconda" (reddish olive).

33—Double; magnitudes 6.0 and 8.0; distance, 2″; close pair.

52—Double; magnitudes 5.7 and 6; distance, 1″.7; close pair.

14—Binary; magnitudes 6 and 7; distance, 1″ in 1880; close pair, in rapid orbital motion; revolution about 250 years.

η (*Eta*)—Double; magnitudes 3.5 and 5; distance, 1″; too close for common telescope.

φ² (*Phi* 2)—Double; magnitudes 5.0 and 11; distance, 2″.8; difficult pair.

32—Binary; magnitudes 4.8 and 7; distance, in 1880, 0″.4; in 1780 the distance was 1″.5; very close pair.

31—Is a variable from 4.7 to 7th magnitude, with a companion of the 11th magnitude at 13″; this star is orange, and its spectrum indicates a sun *beginning to cool off*.

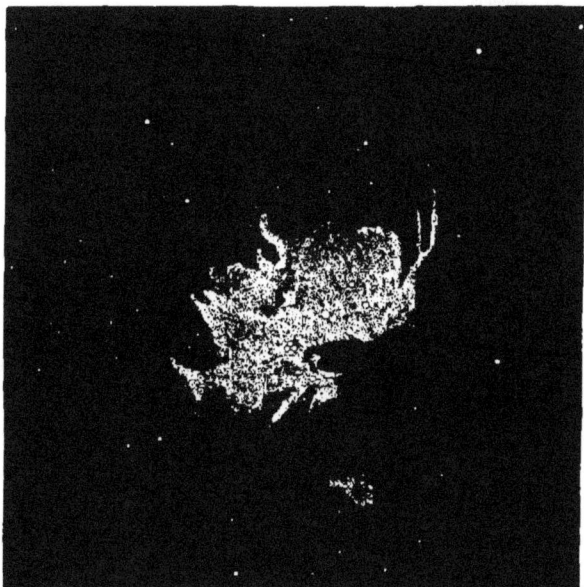

Fig. 127.—Nebula Messier 42- (From a direct photograph taken by Mr. Common.)

M. 42—Is *the finest nebula of all the heavens.* Mr. Bond has written a book on this nebula alone; the spectroscopical observations prove that it is composed of incandescent gases, probably hydrogen and nitrogen; Mr. Bond thought that it was composed entirely of small stars, but it is not so. Mr. Draper, in the United States, and Mr. Common, in England, have succeeded in photographing it; it took 36 minutes of exposure; a star of the 1st magnitude requires only one-hundredth of a second to be photographed. Its spectrum examination indicates that it is traveling at the rate of 17 miles per second from us, or we from it. The nebula proper covers a space equal to the apparent size of the moon, but the nebulosity extends a great deal more; Secchi followed it for a distance of 4 degrees from east to west, and 5 degrees from north to south. If this nebula was only as far from us as 61 of Cygnus, the nearest star that we can see with the naked eye in our latitudes, it would be at least three trillions of miles long, and a fast train going at the rate of 60 miles an hour would have to keep going for more than 5,650,000 years to traverse it (Flam., Les Etoiles, page 466). The stars θ (*theta*) are in the center of the nebula. (Fig. 127.)

H. VII, 4—Is a fine cluster of some 600 stars, among them a nice pair of 8th and 9th magnitudes; distance 23″ (marked on our planisphere north of 15).

M. 78—Is a quadruple nebula 9 minutes long, 5 minutes wide (see between ζ (*zeta*) and 56).

π6 (*Pi*6)—Southeast of π6 (*pi*6) there are five or six red stars; some of them very red; visible in a common telescope.

MONOCEROS.

Monoceros, or the Unicorn, first appears on the planisphere of Bartschius, in 1624, but it was already known a little before, and mention of it is made in a book published in Frankfort in 1564, under the name of Neper (the Forest). In the Persian sphere, brought up by Scaliger, this fantastic animal is also found.

DESIGNA- TION.	MAGNI- TUDE.	POSITION		DESIGNA- TION.	MAGNI- TUDE.	POSITION	
		R. A. h. m.	1880 DECL. ° ′			R. A. h. m.	1880 DECL. ° ′
30	4.0	8.20	− 3.31	2	5.7	5.53	− 9.34
11 trip.	4.2	6.23	− 6.57	12176 Lal.	5.8	6.16	−11.43
26	4.2	7.36	− 9.16	7	5.9	6.14	− 7.46
5 dbl.	4.4 org.	6. 9	− 6.14	P. VII, 228	6.0	7.45	− 8.51
22	4.5	7. 6	− 0.17	12	6.0	6.26	+ 4.57
8 dbl.	4.7	6.17	+ 4.39	P. VI, 82	6.5	6.17	+ 3.49
31	4.9	8.38	− 6.40	15 S, trip.	4.V.	6.34	+10. 0
13	5.0	6.26	+ 7.26	T	6.V.	6.19	+ 7. 9
29 trip.	5.0	8. 3	− 2.38	U	6.V.	7.25	− 9.32
18	5.2	6.42	+ 2.33	W. B 669	5.V.	7.23	− 1.39
28	5.3	7.55	− 1. 3	*	6.0 red	6.36	− 9. 3
10 dbl.	5.4 yel.	6.22	− 4.42	*	7.5 red	7.37	−10.36
17	5.4	6.41	+ 8.10	*	7.0 org.	6.24	− 2.57
12494 Lal.	5.5	6.25	+11.38	*	6.0 org.	7.23	−10. 5
20	5.5	7. 4	− 4. 3	H. VII, 2	cl.	6.26	+ 4.57
19	5.6	6.57	− 4. 4	H. IV, 2	neb.	6.33	+ 8.52
3 dbl.	5.6	5.56	−10.36	H. VII, 35	cl.	6.21	+12.42
27	5.6	7.54	− 3.21	M. 50	cl.	6.57	− 8.10
25	5.7	7.31	− 3.50	H. VI, 22	cl.	8. 7	− 5.26
12587 Lal.	5.7	6.28	+ 7.40	H. VI, 27	cl.	6.46	+ 0.36

NOTES.

11—Triple; magnitudes 5–5.5 and 6; distances, 7″ and 2″.5; the companions are only 2″.5 apart.

8—Double; magnitudes 4.7 and 7.5; distance, 14″; nice pair; yellow and bluish. The magnitudes of this pair have been noted by Lalande, in January, 1794, 6½ and 7; and in February, 1797, 4 and 8½; by Piazzi, of 5½ and 8; by Struve, in 1822, of 4.5 and 7; in 1831, of 4 and 6.7; they must be variable stars.

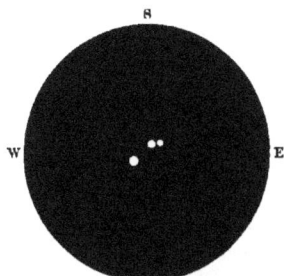

Fig. 128.—Triple Star 11.

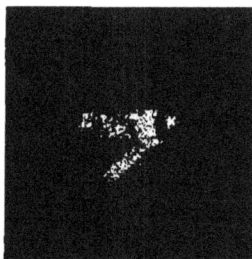

Fig. 129.—Nebula II. IV, 2.

15 S.—Varies irregularly from the 4th to 6th magnitude in a very short period of 3 or 4 days, not well determined; it is also double, which is very rare for a variable; magnitudes 5 and 10; distance, 3″; orange and blue; there is another companion of 13th magnitude 16″ apart, and still another smaller; nice field for small telescope.

29—Triple; magnitudes 5.0-11 and 9; distances, 30″ and 67″; the second one is difficult to be seen, and seems to vary from the 10th to 12th magnitude.

R.—Between 13 and 15; varies from the 9th to the 13th magnitude; too small for common telescope.

T.—Varies from 6.2 to 7.6 in 27 days.

U.—Ten minutes west of 26; varies from 6.2 to 7.6 in 46 days.

H. VII, 2—Is a little "amas" with the reddish star 12 of 6th magnitude in the center (see below 13).

H. IV, 2—Is a triangular nebula like the tail of a comet (see near 15).

M. 50—On a line between Sirius and Procyon there is an "amas" in which can be seen a nice little pair and a red star.

H. VI, 22—South of 29 is a cluster of some fifteen stars of the 9th magnitude, almost visible to the naked eye.

This constellation traversed by the Milky Way is very interesting.

CANIS MINOR.

Canis Minor, the Little Dog, is also a very old constellation; Procyon, its brightest star, is already mentioned by Eudoxus.

DESIGNA-TION.	MAGNI-TUDE.	POSITION		DESIGNA-TION.	MAGNI-TUDE.	POSITION	
		R. A. 1880 h. m.	DECL. ° ′			R. A. 1880 h. m.	DECL. ° ′
α (Procyon)	1.4	7.33	+ 5.32	6	4.8	7.23	+12.15
β trip.	3.0	7.21	8.32	11	5.5	7.40	11. 5
γ	5.2	7.22	9.10	P. VII, 289	4.7	7.56	2.50
δ1	5.8	7.26	2.11	P. VII, 249	6.4	7.49	9.11
δ2	6.2	7.27	3.33	Σ 1126 dbl.	7.0	7.35	5.34
ε	5.4	7.19	9.31	R	7.V. red	7. 2	10.13
ζ	5.4	7.45	2. 4	S	7.V. red	7.26	8.34
η	5.9	7.22	7.11	*	7.1 red	7.37	5.14

NOTES.

α (Alpha) Procyon —Is a star in rapid proper motion; 1″.27 S. W. per year; 21 minutes in 1,000 years. If it keeps going in the same direction and at the same rate it will cross the equator 12,000 years from now and will be part of the Southern Hemisphere.

Auwers, in 1861-62, found for the parallax of Procyon..0″.240±0″.029

Wagner, in 1863-83.....................................0″.299±0′.038

Elkin, of Yale College, in 1887-88................ ....0″.266±0″.047

Taking for the average 0″.27, it would be 761,000 times more distant than the earth is from the sun, or 70 trillions of miles, and the light would have to travel 12 years to reach us (Revue d'Ast., 1889; page 450). The motion of Procyon is not regular and does not follow a straight line. Mr. Auwers' observations tend to prove that there is a star near by, forming

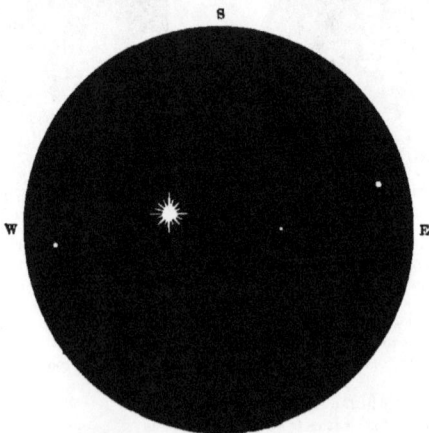

Fig. 130.—Multiple Star Procyon.

a binary system and revolving around their center of gravity in about 40 years. Similar perturbations were noticed for Sirius by Bessel, and its companion was seen 18 years afterward. (See Canis Major notes.)

β (*Beta*) *Gomeisa*—Is a triple star.

Σ 1126—Binary; magnitudes 7.0 and 7.3; distance, 1″.6; very close pair; near Procyon.

CANIS MAJOR.

Canis Major, the Greater Dog, is spoken of in Homer and Hesiod; it is one of Orion's dogs in pursuit of the Hare.

DESIGNA- TION.	MAGNI- TUDE.	POSITION R. A. 1880 h. m.	DECL. ° ′	DESIGNA- TION.	MAGNI- TUDE.	POSITION R. A. 1889 h. m.	DECL. ° ′
α (Sirius) bin.	1.0	6.40	− 16.33	15	5.3	6.48	−20. 4
β	2.2	6.17	17.54	19 dbl.	4.9	6.50	19.59
γ	4.5	6.58	15.27	22	3.6 red	6.57	27.46
δ dbl.	2.1	7. 4	26.12	27	Var.	7. 9	26. 9
ε (Adara)dbl.	1.9	6.54	28.49	28	4.2	7.10	26.34
ζ dbl.	3.2	6.16	30. 1	29	5.6	7.14	24.20
η dbl.	2.9	7.19	29. 4	30 dbl.	4.6	7.14	24.44
θ	4.4 red	6.49	11.53	11985 Lal.	5.5	6.10	13.41
ι	4.9	6.51	16.54	12541 Lal.	5.6	6.26	12.18
κ	4.0	6.45	32.22	2147 B.A.C.	6.0	6 29	31.56
λ	4.7	6.24	32.30	2162 B.A.C.	5.7	6.30	32.37
μ bin. (?)	5.5 red	6.51	13.53	12825 Lal.	5.3	6.34	14. 2
6 ν¹ dbl.	6.4	6.31	18.34	2291 B.A.C.	6.0	6.54	25.15
7 ν²	4.2	6.31	19. 9	12278 Lal.	5.6	6.19	11.28
8 ν³	4.9	6.33	18. 8	13059 Lal.	5.7	6.40	14.40
4 ξ¹	4.5	6.27	23.20	14200 Lal.	5.3	7.12	23. 6
5 ξ²	4.8	6.30	22.52	2244 B.A.C.	7.0	6.45	27.12
16 o¹	3.9 red	6.49	24. 2	*	8.1 very red	6.19	26.59
24 o²	3.4	6.58	23.39	*	7.5 red	7. 2	11.44
				M. 41	cl.	6.42	20.37
10 dbl.	5.7	6.40	30.57	H. VII, 12	cl.	7.12	15.25
11 trip.	5.5	6.41	−14.18	H. VII, 17	cl.	7.14	24.44

NOTES.

Sirius, the Dog Star—Is the brightest white star of the heavens; 3,285 years B. C. Sirius was rising at the solstice of summer; its apparition was the signal of the Nile's inundation and regulated the Egyptian calendar, hence its name *Dog Star*. Most of the astronomers believed from "doubtful citations" that the color of Sirius has changed, and was red about 2,000 years ago; this opinion has been discussed by M. Flammarion in his work "Les Etoiles," pages 477 to 479, and he came to the conclusion that "it is possible" but "far from certain." The stars seen through the telescopes do not show *any diameter*; the stronger the power the smaller is the appearance of the star; the instrument only amplifies their brightness. When Sirius enters the field of the telescope it has the appearance of a strong, sudden flash of light. Mr. Wallaston after many experiments declared that the diameter of Sirius is not one-fiftieth of a second; if it is only that diameter it represents 17,000,000 of miles; about 20 times the diameter of the sun, and 7,000 times its volume. Seen from Sirius the sun would be a star of the 6th magnitude only. The light of Sirius is so strong that it can be seen in daytime (when you know its position) with a

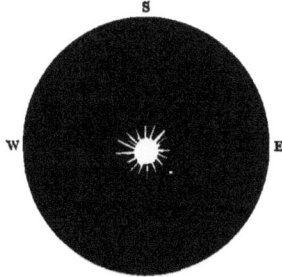

Fig. 131.—Sirius and its companion, in 1880.

small telescope. This beautiful star is moving toward ζ (*zeta*) near the constellation of Columba, and will be close to it in 40,000 years; it comes from the same region that the sun itself occupied many thousand years ago, and they are now going in opposite directions from each other. This motion is not uniform and does not follow a straight line; its position is sometimes east, sometimes west, of the general direction. The period is 49 years; in 1843 Sirius was 0″.152 west of the line and in 1867 it was 0″.152 east of it. In 1844, Bessel suggested that these perturbations were most likely produced by a satellite of Sirius; in 1851 Peters calculated the orbit that would answer the question, and 11 years afterward, the son of Alvan Clark, then 14 years old, in looking through the telescope which his father was making for the Dearborn University of Chicago, saw it and said: "Father, Sirius has a companion." Its position was found to be where the theoretical orbit of Peters would have it. Bessel has been dead since 1846, 16 years before the above discovery was made. Since then the companion of Sirius has been observed many times, but it is only visible in large telescopes and when the atmosphere is pure. Mr. Burnham has tried to see it in 1891, under very favorable circumstances, and has failed to see the least trace of it. The distance, in 1880, was 10″ and, in 1890, 4″.19. This celebrated observer has also drawn the orbit of the companion from a complete series of measures, and this ellipse gives a period of 53 years. In comparing the observations some other perturbations have been noticed, which would indicate the presence of another companion. We give here the parallax of Sirius obtained at different times:

OBSERVERS.	PARALLAX.	OBSERVERS.	PARALLAX.
Henderson, 1832-33	0″.340±0″.25	Belopolsky, 1888, from Wagner's observations, 1862-70..	0″.430±0″.099
Abbe, 1866	0″.270±0″.10	Gill, 1881-83	0″.370±0″.009
Gylden, from Maclear's observations, 1836-37	0″.193±0″.087	Elkin, 1881-83	0″.407±0″.018
		Elkin, 1888	0″.266±0″.047

In taking the average, or 0″.33, it would represent 625,000 times the distance of the earth from the sun, or 58 trillions of miles, and the light would have to travel over 9 years and 10 months to reach us (Revue d'Ast., 1889; page 444).

The spectrum of Sirius (Fig. I, page viii) indicates a high temperature; its photosphere is composed of hydrogen, in which are found in dissociation the vapors of iron, magnesium, sodium, etc.; the lines of hydrogen are very prominent and the lines of the other metals quite feeble. From phometrical observations the diameter of Sirius would be 12 times larger than the diameter of the sun, its surface 144 times, and its volume 1728 times (Flam., Les Etoiles, page 484).

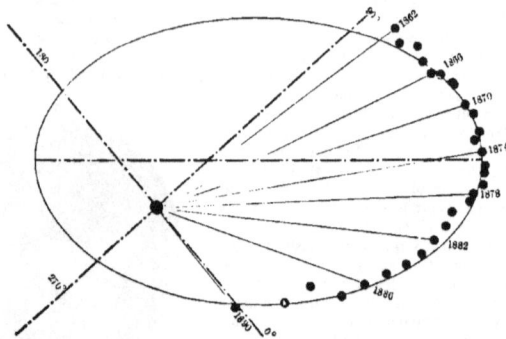

Fig. 132.—Orbit of Sirius. (From a diagram by Mr. Burnham).

Mr. Burnham said: "If this ellipse is correct the minimum distance will be 2".4 in 1892.5, and about the end of 1894 the distance will be the same as at the time of my last measures. At that time, therefore, I trust the large telescope will supply observations which will definitely settle most of the uncertainties in the orbit of this interesting system." (Lick Observatory, February 26, 1891).

β (Beta)—Also called Mirzam; has a little companion of 9th magnitude at 105"; difficult.

δ (Delta)—Double; magnitudes 2.1 and 7.5; distance, 2' 45"; visible with an opera glass.

ζ (Zeta)—Double; magnitudes 3.2 and 7; distance, 2' 47"; visible with an opera glass.

30—Double; magnitudes 4.6 and 9; distance, 1' 25"; nice field.

μ (Mu)—Double; magnitudes 5.5 and 9; distance, 3".

ν1 (Nu1)—Double; magnitudes 6.4 and 8; distance, 17"; easy pair.

17—Quadruple; magnitudes 6-9-10 and 11; distances, 45", 52" and 125"; interesting group.

o1 (Omicron1)—Double; magnitudes 3.4 and 8; distance, 30"; the main star is orange and somewhat variable.

22—Is a very red star; it was noted of 4th magnitude by Hevelius in 1660; Maraldi, in 1670, did not see it, but in 1692 and 1693 noted it of the 4th magnitude; Lalande did not observe it at all; it varies probably from the 3d to the 6th magnitude.

38—Also seems to vary from the 4½th to the 7th magnitude.

M. 41—Near 15 is a nice cluster, sometimes visible to the naked eye, composed of 92 stars of the 8th to the 12th magnitude; easily seen in common telescopes; near the center is a red star brighter than the others.

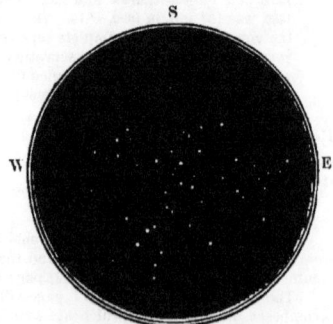

Fig. 133.—Cluster Messier 41.

H. VII, 12—East of γ (gamma) is another rich cluster, composed mostly of stars of the 10th magnitude.

SEXTANS.

Sextans Uraniæ, or the Sextant, is a constellation formed by Hevelius in 1690.

DESIGNA-TION.	MAGNI-TUDE.	R. A. 1880 h. m.	DECL. ° '	DESIGNA-TION.	MAGNI-TUDE.	R. A. 1880 h. m.	DECL. ° '
1	5.4	9.31	+ 7.22	30	5.2	10.24	− 0.59
2	5.2	9.32	+ 5.11	31	7.0	10.24	+ 2.46
8 bin.	5.4	9.47	− 7.32	35 dbl.	6.2	10.37	+ 5.23
12	6.8	9.53	+ 3.57	37	6.0	10.40	+ 7.0
15	4.7	10. 2	+ 0.13	41 dbl.	6.0	10.44	− 8.16
18	6.0 red	10. 5	− 7.49	19662 Lal.	6.3	9.58	− 8.59
19	6.2	10. 7	+ 5.12	19823 Lal.	8.0	10. 5	− 1. 3
27	6.8	10.21	− 3.46	H. I, 3–4	neb.	10. 8	+ 4. 4
29	5.4	10.23	− 2. 7	H. I, 163	neb.	9.50	− 7. 8

NOTES.

8—Binary; magnitudes 5.6 and 6.5; distance, 0″.5; very close pair; revolution, 33 years.
35—Double; magnitudes 6 and 8; distance, 7″; yellow and blue; very nice pair.
H. I, 3—Is a double nebula north of 15.
H. I, 163—Elliptical nebula 35″ wide, 150″ long; in the center is a star of the 10th magnitude; it is between 18 and 19662.

HYDRA.

Hydra, or the Sea Serpent, is the famous hundred-headed monster killed by Hercules on the shores of Lake Lerna.

On the Hydra can be seen a "cup" and a "crow." According to Greek mythology, Apollo sent the crow with a cup to bring some water for a sacrifice to Jupiter, but the crow came across a fig tree and waited for the figs to become ripe; as he was afraid he would be reprimanded for his delay, he blamed the Hydra for it; Apollo punished him by turning his plumage, which was white, into black, and placing him and the cup on the back of the Sea Serpent (Flam., Les Etoiles, page 528). They appear already in Eudoxus' time.

DESIGNA-TION.	MAGNI-TUDE.	R. A. 1880 h. m.	DECL. ° '	DESIGNA-TION.	MAGNI-TUDE.	R. A. 1880 h. m.	DECL. ° '
α dbl.	2.V. org.	9.22	− 8. 8	1	6.2	8.19	− 3.22
β dbl.	4.5	11.47	−33.14	2	6.5	8.20	− 3.36
γ	3.3 org.	13.12	−22 32	12	4.4	8.41	−13. 7
δ dbl.	4.1	8.31	+ 6. 9	14	5.8	8.43	− 2.60
ε bin.	3.5	8.40	+ 6.51	24	6.0	9.11	− 8.14
ζ	3.1	8.49	+ 6.24	25	7.5	9.14	−11.28
η	4.5	8.37	+ 3.49	26	5.4	9.14	−11.28
θ dbl.	3.8	9. 8	+ 2.49	27	5.5	9.15	− 9. 3
ι	4.0 red	9.34	− 0.36	51 dbl.	5.0	14.16	−27.12
κ	5.3	9.35	−13.47	52	4.7	14.21	−23.57
λ	3.4	10. 5	−11.45	54 dbl.	5.2 red	14.39	−24.56
μ dbl.	4.0 yel.	10.20	−16.13	58	4.8	14.43	−27.28
ν	3.2	10.44	−15.34	18639 Lal.	5.2	9.22	−21.49
ξ dbl.	3.8	11.27	−31.12	19034 Lal.	5.3	9.36	−23. 2
ο	5.0	11.34	−34. 5	19093 Lal.	5.5	9.37	−23.23
π	3.6 yel.	14. 0	−26. 6	20556 Lal.	5.V. red	10.32	−12.45
ρ	4.8	8.42	+ 6.17	P. VIII, 167	5.6	8.41	− 1.27
σ	5.0	8.32	+ 3.48	P. X, 256	5.7	11. 3	−27.25
31 τ1 dbl.	4.8	9.23	− 2.14	P. XI, 96 dbl.	5.2	11.26	−28.36
32 τ2	4.8	9.26	− 0.39	P.VIII,108 dbl.	6.0	8.30	+ 7. 3
39 υ1	4.1	9.46	−14.17	R	4.V. red	13.23	−22.40
40 υ2	4.5	9.59	−12.29	S	8.V. org.	8.47	+ 3.31
φ	5.0	10.33	−16.15	T	7.V. red	8.50	− 8.41
χ	4.8	10.59	−26.38	*	6.0 red .	10.46	−20.36
ψ	5.4	13. 3	−22.28	*	7.0 red	13.42	−27.46
ω	5.5	9. 0	+ 5.35	*	6.5 org.	9.46	−22.27
				*	7.5 org.	9.14	+ 0.41
33 A	6.0	9.29	− 5.22	H. IV, 27	neb.	10.19	−18. 2
b1	5.8	10.41	−16.40	M. 68	cl.	12.33	−26. 5
b2	5.5	10.45	−17.41				

NOTES.

α (Alpha)—Also called Alphard from the Arab Al-fard (the solitary), now orange; was noted red by Sufi; the Chinese called it "the red bird;" it seems to vary from the 1st to 2d magnitude, and has most likely changed its color; it is also double. It was observed as passing the meridian at sunset the day of the vernal equinox during the time of the Chinese Emperor Yao, about 2,350 years before Christ; it is one of the oldest astronomical observations (Flam., Les Étoiles, page 530).

ε (Epsilon)—Binary; magnitudes 3.6 and 7.5; distance, in 1880, 3".5; beautiful pair; yellow and blue; revolution about 700 years. Schiaparelli finds the large star a close pair; distance, 0".2; the close pair is in rapid orbital motion.

54—Double; magnitudes 5.2 and 8; distance, 9"; elegant pair; yellow and violet.

τ (Tau)—Double; magnitudes 4.8 and 8; distance, 65"; very easy pair.

P. XI, 96—Double; magnitudes 5.2 and 6.5; distance, 10"; bright pair.

P. VIII, 108—Double; magnitudes 6 and 7; distance, 10"; nice pair in a fine field.

H. IV, 27—South of μ (mu) is a curious nebula said by Admiral Smyth to look like Jupiter for its size, its light and its color; in its center shines a nice little star, and, besides, four little stars appear in the field; the spectroscopical analysis indicates that it is entirely gaseous.

M. 68—South of β (beta) of Corvus is a cluster 4' long by 3' wide; quite pale; between two little stars.

R.—There are several variables in this constellation; the principal one is R, which varies from the 4th to the 10th magnitude in a period growing shorter every year; the results of its observations are very interesting. Hevelius, at Dantzic, saw it of 6th magnitude in 1662; Montanari, at Bologna, saw it of 4th magnitude in 1670; Maraldi, at Paris, saw it of the 4th in 1704 and fol-

Fig. 134.—Elliptic Nebula **H. IV, 27** and Companions.

lowed it until 1712; after that it was lost and nearly forgotten, but Pigott, at York, observed it again in 1784; in the 19th century Argelander and Schmidt paid special attention to it; comparing the different observations it seems that the period of variability diminished from 9 to 10 hours for every revolution; it was 547 days in 1680, 487 days in 1780, 432 days in 1880. What will it be in 1980? The other variables are too small for common telescopes.

CRATER.

Crater, or the Cup, is one of the forty-eight constellations of the ancient Greeks. (See Hydra.)

DESIGNA-TION.	MAGNI-TUDE.	POSITION		DESIGNA-TION.	MAGNI-TUDE.	POSITION	
		R. A. 1880 h. m.	DECL. ° '			R. A. 1880 h. m.	DECL. ° '
α mult.	4.4	10.54	−17.40	ι	5.8	11.33	−12.33
β	4.6	11. 6	22.10	κ	6.1	11.21	11.42
γ dbl.	4.2	11.19	17. 2	λ	5.4	11.17	18. 7
δ dbl.	3.5	11.13	14. 7	31	5.5	11.55	19. 0
ε	5.5	11.19	10.12	21203 Lal.	5.7	10.57	10.39
ζ	5.2	11.39	17.41	R	8.V. red	10.55	17.41
η	5.4	11.50	16.29	*	6.0 org.	10.53	15.42
θ	5.0	11.31	9. 8				

NOTES.

α (Alpha)—Also called Alkes, has several small companions; outside of the variable R there is nothing worth mentioning in this small constellation.

R.—Very near α (alpha) is a star of a fiery red, varying from the 8th to the 10th magnitude in periods alternatively of 72 days and 88 days.

CORVUS.

Corvus, or the Crow, is a very old constellation. (See Hydra.)

DESIGNA-TION.	MAGNI-TUDE.	R. A. 1880 h. m.	DECL. ° '	DESIGNA-TION.	MAGNI-TUDE.	R. A. 1880 h. m.	DECL. ° '
α	4.2 red	12. 2	−24. 4	η	4.5	12.26	−15.32
β dbl.	2.6 red	12.28	22.44	P. XII, 54	5.6	12.15	12.54
γ	2.V.	12.10	16.52	23675 Lal. dbl.	5.8	12.35	12.21
δ dbl.	3.V.	12.24	15.51	23726 Lal.	7.5	12.37	13.12
ε	3.3 yel.	12. 4	21.57	Σ 1664 dbl.	7.5 red	12.32	10.49
ζ	5.2	12.14	21.33	It	7.V. red	12.13	18.35

NOTES.

α (*Alpha*)—Also called Alchiba, is not the brightest star of this constellation; it is γ (*gamma*), which is now of the 2d magnitude, and noted only of 4th by Bayer in 1603; at his time α (*alpha*), β (*beta*), γ (*gamma*) and δ (*delta*) *Algores* were all of the 4th magnitude; Alpha is now the 5th in brilliancy, as our catalogue and planisphere plainly show; α (*alpha*) and β (*beta*) are red.

δ (*Delta*)—Double; magnitudes 3.0 and 9; distance, 24″; the companion is quite dark and difficult. 23675—Double; magnitudes 6.4 and 6.5; distance, 5″.8; nice pair.

R.—2° S. E. of γ (*gamma*); varies from the 7th to 12th magnitude in 318 days.

PISCIS NOTIUS.

Piscis Notius, or the Southern Fish is one of the forty-eight constellations of the ancients, and appeared already on the sphere of Eudoxus.

DESIGNA-TION.	MAGNI-TUDE.	R. A. 1880 h. m.	DECL. ° '	DESIGNA-TION.	MAGNI-TUDE.	R. A. 1880 h. m.	DECL. ° '
α	1.7	22.51	−30.15	θ dbl.	5.2	21.41	−31.27
β dbl.	4.4	22.25	32.58	ι	4.4	21.38	33.34
γ dbl.	4.6	22.46	33.31	λ	5.6	22. 7	28.21
δ dbl.	4.4 red	22.49	33.11	μ	4.7	22. 1	33.34
ε	4.3	22.34	27.40	P. XXI, 46	4.9	21.11	32.40
ζ	6.7	22.24	26.41	9350 Lac.	5.3	22.57	35.24
η dbl.	5.7	21.54	29. 2	9352 Lac.	7.5	22.57	36.27

NOTES.

α (*Alpha*) Fomalhaut—Is the brightest star of this constellation.
κ (*Kappa*)—Noted by Bayer; does not exist.
9352 Lac.—Is a small star of the 7½th magnitude; in rapid motion; 6″.96 for one year; the greatest after 1830 Groombridge (see our Notes of Ursa Major); motion discovered by Prof. Gould, at Cordoba, Argentine Republic, in 1880–1. Mr. Gill, in 1883, gave for its parallax 0″.285±0″.020; if correct, it is one of the nearest stars to us, 711,000 times the distance of the earth from the sun, or 65 trillions of miles; time for the light to reach us is 11 years and 88 days.

APPARATUS SCULPTORIS.

Apparatus Sculptoris, or the Sculptor's Shop, is a constellation, first introduced by Lacaille in 1752.

DESIGNA-TION.	MAGNI-TUDE.	R. A. 1880 h. m.	DECL. ° '	DESIGNA-TION.	MAGNI-TUDE.	R. A. 1880 h. m.	DECL. ° '
α P. o, 250	4.2	0.53	−30. 0	κ² P. o, 6	5.2	0. 5	−28.28
β 9513 Lac.	6.7	23.26	26.24	η P. o, 79	5.2	0.22	33.40
γ P. XXIII, 36	4.4	23.12	33.11	ε P. I, 168 dbl.	5.4	1.40	25.39
δ P. XXIII,				158 Lal.	5.4	0. 8	8.30
192 dbl.	4.6	23.43	28.48	P. o, 111 dbl.	6.5	0.28	35.39
ζ P. XXIII, 259	5.2	23.56	30.23	*	6.0 red	23.51	27.18
κ¹ 9741 Lac.	5.5	0. 3	28.39	*	6.0 org.	1.21	33.10

NOTE.

This constellation, entirely visible from our latitude, contains only three stars above the 5th magnitude.

CETUS.

Cetus, or the Sea Monster, is mentioned by Eudoxus, Aratus, Hipparchus and Ptolemy, who called it "Ketos" (the Whale); Hyginus named it "Orphos," a fish quite different to the whale; according to others, it is the sea monster sent by Neptune to ravage the shores of Ethiopia and kill Andromeda.

DESIGNA-TION.	MAGNI-TUDE.	POSITION		DESIGNA-TION.	MAGNI-TUDE.	POSITION	
		R. A. 1880 h. m.	DECL. ° ′			R. A. 1880 h. m.	DECL. ° ′
α dbl.	2.4 org.	2.56	+ 3.37	2	4.3	23.58	—18. 0
β dbl.	2.2 yel.	0.38	—18.39	3	5.2	23.58	—11.11
γ dbl.	3.2	2.37	+ 2.44	6	5.1	0. 5	—16. 7
δ	4.0	2.34	— 0.11	72 Lal.	5.4	0. 6	—18.36
ε	4.5	2.34	—12.23	7	4.3	0. 9	—19.36
ζ dbl.	3.5	1.46	—10.55	P. 0. 91	5.2	0.24	—24.27
η dbl.	3.5 org.	1. 3	—10.49	12 trip.	6.0	0.24	— 4.37
θ	3.2	1.18	— 8.48	13	6.0	0.29	— 4.15
ι dbl.	3.5 yel.	0.13	— 9.31	20	5.2	0.47	— 1.47
96 κ¹	5.1	3.13	+ 2.55	37 qdl.	5.3	1. 8	— 8.34
97 κ²	6.2	3.15	+ 3.14	42 bin.	6.0	1.14	— 1. 8
λ	4.7	2.53	+ 8.25	46	5.1	1.20	—15.13
μ	4.2	2.38	+ 9.36	48	5.3	1.24	—22.15
ν dbl.	5.0	2.30	+ 5. 4	3139 Lal.	5.2	1.37	— 4.19
65 ξ¹	4.3	2. 7	+ 8.17	56	5.0	1.51	—23. 7
73 ξ²	4.2	2.22	+ 7.55	61 qdl.	6.5	1.58	— 0.55
o dbl.	2.V.	2.13	— 3.31	66 dbl.	6.0	2. 7	— 2.58
π	4.0	2.38	—14.22	94 dbl.	5.3	3. 7	— 1.39
ρ	4.6	2.20	—12.50	Σ 101, dbl.	8.0	1. 8	— 8.17
σ	4.7	2.26	—15.46	Σ 106, dbl.	8.5	1.10	— 7.48
τ	3.4	1.39	— 16.34	Σ 147, dbl.	6.0	1.34	—11.54
ν	4.0	1.54	—21.40	Σ 218, dbl.	7.0	2. 3	— 1. 1
17 φ¹	5.1	0.38	—11.15	S	7.V. org.	0.18	—10. 0
19 φ²	5.5	0.44	—11.17	R	8.V. org.	2.20	— 0.43
22 φ³	5.7	0.50	—11.55	2598 Lal.	6.V.	1.20	— 4.35
23 φ⁴	5.9	0.53	— 12. 1				
χ dbl.	4.8	1.44	—11.17				

NOTES.

α (*Alpha*)—Is also called *Menkar*, β (*beta*) *Diphda*, and ζ (*zeta*) *Baten Kaitos.*

o (*Omicron*) *or Mira*—Is a famous variable; the first noticed by David Fabricius, in August, 1596; the 6th of November, 1779, Mira was nearly as bright as Aldebaran; sometimes it comes up to the 1st magnitude and sometimes it stops at the 4th magnitude; in a period of about 166 days it came down below the 9th magnitude; the time of the variability seems to be 331 days 8 hours and 4 minutes. At each period it remains about 5 months invisible to the naked eye, its magnitude being lower than the 6th; then it becomes gradually brighter, and visible to the naked eye for about three months, rising as we have before mentioned to the 2d magnitude; but its maximum lasts no more than 15 days. Its spectrum seems to indicate that it is sometimes covered with spots like our sun, but the period, instead of being about 11 years, is only 11 months. Mira has a companion of 9.5 magnitude: distance, 1′ 58″.

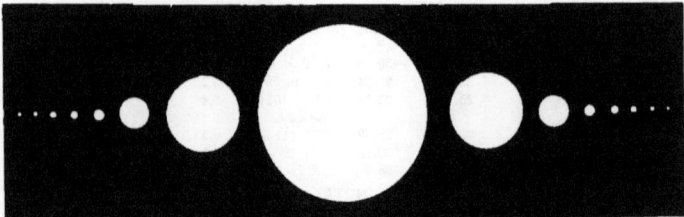

Fig. 135.—Diagram Showing the Variations of Mira.

ζ *(Zeta)*—Double; magnitudes 3.5 and 9; distance, 2′ 45″; easy pair.

χ *(Chi)*—Double; magnitudes 4.8 and 7.5; distance, 3′ 6″; easy pair.

γ *(Gamma)*—Double; magnitudes 3.2 and 7; distance, 3″; pale yellow and blue; nice contrast.

37—Double; magnitudes 5.3 and 7; distance, 51″; in the field will be found a nice pair, magnitudes 8 and 10; distance, 20″; yellow and violet; nice field of small stars.

66—Double; magnitudes 6.5 and 8; distance, 15″; yellow and blue; elegant pair.

Σ 147—Double; magnitudes 6 and 7; distance, 3″.5; nice pair.

ν *(Nu)*—Double; magnitudes 5.0 and 11; distance, 6″; difficult pair.

61—Double; magnitudes 6.5 and 11; distance, 39″; difficult pair.

Σ 218—Double; near 61; magnitudes 7 and 8.5; distance, 4″.6; nice field.

84—Double; magnitudes 7.5 and 10; distance, 4″.7; difficult pair; companion lilac.

42—Binary; magnitudes 6 and 7½; distance, 1″.4; very slow orbital motion.

τ *(Tau)*—Is in rapid motion, 3′ 20″ in 100 years; will be near η *(eta)* 19,000 years from now (Flam., Les Etoiles, page 502).

Near δ *(delta)* there is a stellar nebula, M. 77, easy with an ordinary telescope.

APPARATUS CHEMICUS or FORNAX.

Apparatus Chemicus, or Chemical Apparatus, is a modern constellation, having been formed by Lacaille, in 1752.

DESIGNA-TION.	MAGNI-TUDE.	POSITION			DESIGNA-TION.	MAGNI-TUDE.	POSITION		
		R. A. h. m.	1880	DECL. ° ′			R. A. h. m.	1880	DECL. ° ′
P. III, 13	3.6	3. 7		−29.28	P. II, 73	5.6	2.17		−24.22
P. II, 195	4.5	2.45		32.54	P. II, 122 dbl.	4.8	2.29		28.46
P. I, 168	5.3	1.40		25.45	P. II, 200	5.6	2.45		28.26
P. I, 241	5.5	1.56		30.36	P. III, 142	4.9	3.38		32.19
P. I, 251	4.8	1.59		29.54	P. III, 176	5.6	3.43		30.31
P. II, 28	5.4	2. 8		31.18	P. II, 194 dbl.	6.5	2.43		37.55

NOTE.

This constellation, entirely visible from our latitude, contains only one star above the 4th magnitude.

ERIDANUS.

Eridanus, the River, also called the River Po, was called at the time of Eudoxus Orion's River, or Potamos (the River); it is the river in which Phaeton, son of Phœbus (the Sun) and Clymene, was drowned after trying to drive the chariot of his father; according to mythology, Phaeton, unable to manage the fiery steeds, was precipitated into the river by Jupiter to prevent a general conflagration.

DESIGNA-TION.	MAGNI-TUDE.	POSITION			DESIGNA-TION.	MAGNI-TUDE.	POSITION		
		R. A. h. m.	1880	DECL. ° ′			R. A. h. m.	1880	DECL. ° ′
α (Achernar)	1.6 red	1.33		−57.51	10 ρ2 dbl.	5.3	2.57		− 8.10
β dbl.	2.8	5. 2		5.15	4969 Lal.	5.7	2.34		9.58
γ dbl.	2.8	3.52		13.51	1 τ1	4.5	2.39		19. 5
δ	3.3	3.38		10.10	2 τ2	4.9	2.46		21.30
ε	3.6	3.27		9.52	11 τ3	4.1	2.57		24. 6
ζ	4.9	3.10		0.16	16 τ4 dbl.	3.4	3.14		22.12
η dbl.	3.7	2.51		0.22	19 τ5	4 5	3.29		22. 2
θ dbl.	2.6	2.54		40.47	27 τ6	3.9	3.42		23.36
ι	4.2	2.30		40.22	28 τ7	5.5	3.43		24.15
κ	4.2	2.23		48.14	33 τ8	4.4	3.49		24.58
λ	4.0	5. 3		8.55	36 τ9	4.4	3.55		24.21
μ	4.0	4.40		3.28	50 υ1	4.7	4.29		30. 0
ν	3.8	4.30		3.36	52 υ2	3.7	4.31		30.49
ξ	5.6	4.18		4. 1	43 υ3	4.0	4.20		34.18
38 o1	4.0	4. 6		7. 9	41 υ4	3.3	4.13		34. 6
40 o2 trin.	4.4	4.10		7.47	φ dbl.	3.5	2.12		52. 2
π	4.7 org.	3.40		12.29	χ dbl.	3.9	1.51		52.14
9 ρ1	5.6 org.	2.55		8. 9	ψ	5.3	4.56		7.21

ERIDANUS—CONTINUED.

DESIGNA-TION.	MAGNI-TUDE.	POSITION			DESIGNA-TION.	MAGNI-TUDE.	POSITION	
		R. A. 1880 h. m.	DECL. ° '				R. A. 1880 h. m.	DECL. ° '
ω	4.7	4.47	—5.39		53	4.1	4.33	—14.32
39 A	5.2	4. 9	10.33		54	4.6	4.35	19.54
62 b, dbl.	5.9	4.50	3.22		55 dbl.	6.5	4.38	9. 1
51 c	5.8	4.32	2.43		60	5.0	4.45	16.26
4	5.7	2.52	24.21		64	4.8	4.54	12.42
5	5.4	2.54	2.57		P. III, 251	5.8	4. 1	27.59
15	5.3	3.13	22.58		P. IV, 154	5.2	4.34	12.21
17	4.7	3.25	5.29		9284 Lal.	5.V.	4.50	16.37
20	5.3	3.31	17.52		P. III, 88	4.V.	3 26	41.49
32 dbl.	4.7	3.48	3.19		*	6.7 org.	4.28	11. 2
35	5.3	3.55	1.53		H. IV, 26	neb.	4. 9	13. 3
45	5.4	4.26	0.18					

NOTES.

This constellation is not entirely visible from our country, and its brightest star, α (alpha) or Achernar, which means " end of the river," will be found on our little map of the Southern Constellations. (Page 66.) β (Beta) is also called Cursa and γ (gamma) Zaurac.

32 – Double; magnitude 4.7 and 7; distance, 6".7; very nice pair; topaz and marine blue; beautiful colors.

o² (Omicron2) – Trinary; magnitudes 4.4–9.5 and 10.5; distances, for 1885, 82" and 4"; the companions are only 4" apart, and revolve around each other in 139 years. This beautiful system has been observed for parallax

    By Gill, in 1883...0".166±0".018
    By Asaph Hall, in 1883...0".223±0".020

Taking 0".19 for the average, this trinary would be 1,086,000 times the distance from the earth to the sun, or 100 trillions of miles, and the light would have to travel more than 17 years to reach us (Revue d'Ast., 1889; page 446). It has " one of the greatest proper motion," 4".10 per year, apparent diameter of the moon in 500 years; direction S. W., toward γ (gamma), near which it will be in 9,000 years; it was near ξ (zi) 5,000 years ago.

o¹ (Omicron1) – Has no perceptible proper motion.

39 A – Double; magnitudes 5.2 and 9; distance, 6".4; beautiful pair; yellow and blue.

62 b – Double; magnitudes 5.9 and 8; distance, 64"; easy pair.

55 – Double; magnitudes 6.5 and 7; distance, 10"; nice pair. The star 56, at 20 minutes from 55, also appears in the field.

θ (Theta) – Is also a double star, and was noted of 2.6 magnitude in 1873; of 2.8 in 1871; of 3d in 1870, and of 3.3 in 1862.

H. IV, 26 – South of 39 A is a very bright circular nebula looking like a star not exactly at the focus of the telescope.

Many stars of this constellation offer notable variations of magnitude.

LEPUS.

Lepus, or the Hare, is also one of the forty-eight constellations of the ancient Greeks, who used to call it " Lagos; " the Latins named it " Lepus," the Arabs " Al-arnab; " all of which mean the same thing.

DESIGNA-TION.	MAGNI-TUDE.	POSITION			DESIGNA-TION.	MAGNI-TUDE.	POSITION	
		R. A. 1880 h. m.	DECL. ° '				R. A. 1880 h. m.	DECL. ° '
α dbl.	2.7	5.27	—17.54		μ	3.4	5. 8	—16.21
β dbl.	2.9	5.23	20.51		ν	5.7	5.14	12.26
γ trip.	3.5	5.40	22.29		17	5.5	6. 0	16.29
δ	3.7	5.46	20.53		P. IV, 285	5.5	4.56	20.14
ε	3.1 red	5. 0	24.32		P. IV, 289	5.4	4.58	26.27
ζ	3.6	5.42	14.52		P. V, 35	5.4	5.11	27. 4
η	3.8	5.51	14.12		P. V, 70	5.4	5.17	24.53
θ	5.2	6. 1	14.56		10063 Lal.	4.9	5.15	21.22
ι dbl.	4.4	5. 7	12. 1		R	6.V. red	4.54	14.59
κ dbl.	4.2	5. 8	13. 5		M. 79	neb.	5.19	24.38
λ	4.1	5.14	13.18		*	cl.	4.55	13.39

NOTES.

α (Alpha)—Is also called Arneb.

R.—The principal curiosity of this constellation is the variable R, visible in a small telescope below 64 on the line passing through *α (alpha)* and *μ (mu)*. Hind, who discovered it in 1845, said that it is "of the most intense crimson, resembling a drop of blood on the black ground of the sky." It is not visible to the naked eye, and varies from the 6¼th to 8¼th magnitude in a period of 438 days, which also varies. At 1° 4' south of R there is a nice field of small stars.

γ (Gamma)—Double; magnitudes 3.5 and 6.5; distance, 1' 33"; easy pair; triple in large telescopes.

κ (Kappa)—Double; magnitudes 4.2 and 8.5; distance, 3".7; nice pair.

ι (Iota)—Double; magnitudes 4.4 and 12; distance, 13"; companion too small for common telescope.

β (Beta)—Double; magnitudes 2.9 and 11; distance, 3"; too close, and companion too small for ordinary telescopes.

COLOMBA.

Colomba, or the Dove, is a constellation generally attributed to Augustin Royer, in 1679; some said it was introduced by Bartschius in 1624, but it already appears in Bayer's Atlas in 1603.

DESIGNA-TION.	MAGNI-TUDE	POSITION		DESIGNA-TION.	MAGNI-TUDE.	POSITION	
		R. A. h. m.	1880 DECL. ° '			R. A. h. m.	1880 DECL. ° '
α (Phact)	2.5	5.35	—34. 8	ν1	6.4	5.33	—27.57
β	2.9	5.47	35.49	ν2	5.3	5.33	28.46
γ dbl.	4.5	5.53	35.18	ξ	5.4	5.51	37. 8
δ	3.9	6.18	33.23	ο	5.1	5.13	35. 1
ε	4.1	5.27	35.34	π1	6.V.	6. 3	42.17
η	4.0 red	5.56	42.49	π2	5.V.	6. 4	42.08
θ	5.3	6. 4	37.14	σ	5.6	5.52	31.24
κ	4.8	6.12	35. 6	τ	6.4	5.50	29.10
λ	5.2	5.49	33.50	2228 Lac.	6.3	6.15	34.21
μ	5.4	5.42	32.21	2234 Lac.	6.0	6.16	34. 5

NOTES.

In this constellation ζ *(zeta)*, ι *(iota)* and ρ *(rho)* have no star corresponding to them: they do not appear in any catalogue or map.

δ (Delta)—Varies from the 4th to 5th magnitude.

η (Eta)—Also varies from the 4th to the 5th magnitude; this star is generally inserted in Argo Navis. As there is another star in this constellation having the same letter, it may cause confusion; η *(eta)* of Colomba is very red.

π2 (Pi 2)—Varies from 4½th to 6th magnitude; π1 *(pi 1)* varies also, and these two stars are sometimes seen of same magnitude.

ANTLIA PNEUMATICA.

Antlia Pneumatica, or Air Pump, is a new constellation formed by Lacaille, in 1752.

DESIGNA-TION.	MAGNI-TUDE.	POSITION		DESIGNA-TION.	MAGNI-TUDE.	POSITION	
		R. A. h. m.	1880 DECL. ° '			R. A. h. m.	1880 DECL. ° '
α	4.4	10.22	—30.27	ζ2	6.3	9.26	—31.20
β	6.0	11. 4	31.42	η dbl.	5.6	9.54	35.19
γ	7.2	10.19	29. 3	θ	5.2	9.39	27.13
P. X, 66	5.7	10.19	37.24	ι	5.1	10.51	36.29
δ dbl.	6.0	10.24	30. 0	*	7.0 red	10. 7	34.44
ε	5.0 red	9.24	35.25	*	6.5 org.	10.30	38.57
ζ1	6.1	9.26	31.22				

NOTES.

This constellation was formed by Lacaille by taking some isolated stars between Argo Navis and Hydra; the brightest does not attain the 4th magnitude. There have been some changes for most of them since 1752; for example, ε *(epsilon)* is now of the 5th magnitude and should be immediately after *α (alpha)*; it was of the 6th magnitude then; γ *(gamma)* was of the 5th magnitude then and is now invisible to the naked eye.

ARGO NAVIS.

Argo Navis, or the Ship, is the ship used by Jason and the Argonauts to go to the discovery of the "Golden Fleece." This constellation is the largest of all and one of the oldest, appearing in the Greek sphere during Eudoxus' time, 4th century B. C.

PUPPIS.

DESIGNA-TION.	MAGNI-TUDE.	POSITION R. A. 1880 h. m.	DECL. ° '	DESIGNA-TION.	MAGNI-TUDE.	POSITION R. A. 1880 h. m.	DECL. ° '
a (Canopus)	1.0	6.21	−52.38	a	4.0 red	7.48	−40.16
ν	3.5	6.34	43. 5	b	4.9	7.48	38.33
τ	3.2	6.47	50.28	J	4.5	7.50	47.48
L² dbl.	3.V.	7.10	44.26	ζ	2.5	7.59	39.40
π	2.7 red	7.13	36.53	ρ dbl.	3.2	8.02	23.58
σ dbl.	3.5 red	7.26	43. 3	Σ 1120, dbl.	6.5	7.30	14.13
n dbl.	5.7	7.29	23.12	Σ 1121, dbl.	7.2	7.31	14.13
k dbl.	4.5	7.34	26.32	Σ 1138, dbl.	7.0	7.40	14.23
l	4.2 red	7.39	28.40	M. 46	cl.	7.36	14.33
c	3.6 red	7.41	37.41	M. 93	cl.	7.39	23.35
2994 Lac.	3.5	7.44	24.34	H. IV, 39	neb.	7.36	14.27
3001 Lac.	6.V. red	7.44	40.21	H. VIII, 38	cl.	7.31	14.13
P .	4.3	7.46	46. 4				

CARINA.

DESIGNA-TION.	MAGNI-TUDE.	POSITION R. A. 1880 h. m.	DECL. ° '	DESIGNA-TION.	MAGNI-TUDE.	POSITION R. A. 1880 h. m.	DECL. ° '
χ	3.7	7.54	−52.39	q	3.3	10.13	−60.44
ε	2.1	8.20	59. 7	I	4.3	10.22	73.25
d	4.7	8.38	59.20	t¹	5.V. red	10.32	58.56
c	4.0	8.52	60.11	t² dbl.	5.2 red	10.34	58.33
G	4.8	9. 5	72. 7	s	4.6	10.24	58. 7
a	3.8	9. 8	58.28	p	3.6	10.28	61. 4
i	4.3	9. 9	61.49	θ	2.9	10.39	63.46
β	2.0	9.12	69.13	η	1.V.	10.40	59. 3
g	4.8	9.13	57. 2	u	4.1 red	10.49	58.13
ι .	2.5	9.14	58.46	x	4.6	11. 4	58.19
l	4.V.	9.42	61.57	R	4.V.	9.29	62.16
υ	3.3	9.44	64.31	neb. of η		10.40	59. 3
ω	3.6	10.11	69.26	cl. and *	red	7.57	60.30

VELA.

DESIGNA-TION.	MAGNI-TUDE.	POSITION R. A. 1880 h. m.	DECL. ° '	DESIGNA-TION.	MAGNI-TUDE.	POSITION R. A. 1880 h. m.	DECL. ° '
γ dbl.	3.0	8. 6	−46.59	λ	2.5	9. 4	−42.57
e	4.6	8.34	42.34	κ	2.7	9.19	54.30
b dbl.	4.1	8.37	46.13	ψ	3.7	9.26	39.56
o	4.0	8.37	52.30	N	3.2 V.	9.28	56.30
d	4.4	8.40	42.12	φ	3.9	9.53	54. 0
δ dbl.	2.2	8.42	54.16	q	4.0	10.10	41.31
a	4.1	8.42	45.36	p	4.1	10.32	47.36
c	4.6	9. 0	46.37	μ	2.9	10.42	48.47

MALUS.

DESIGNA-TION.	MAGNI-TUDE.	POSITION R. A. 1880 h. m.	DECL. ° '	DESIGNA-TION.	MAGNI-TUDE.	POSITION R. A. 1880 h. m.	DECL. ° '
b	4.4 red	8.35	−34.53	c	4.4	8.45	−27.16
a	3.8	8.39	32.45	e dbl.	4.5	9.05	29.58

NOTES.

This is the largest of all constellations, but we can see only a part of it, the rest being too far in the Southern Hemisphere. (Refer to our map of the Southern Constellations, page 66, for everything that does not appear on our planisphere.)

H. VIII, 38—The second nebula in the direction of Sirius to γ (*gamma*) Canis Majoris, is a nice cluster; visible to the naked eye, with two double stars in the field.

M. 46—Is a circular cluster composed of small stars shining like diamond dust, near which appears the remarkable planetary nebula II. IV, 39. (See near 38.)

ν (*Nu*)—Is a beautiful blue star; color very rare in single stars.

α (*Alpha*) *Canopus*—The brightest star after Sirius; is too far south to appear on our planisphere; it could be seen from all points south of the 37th degree of latitude.

η (*Eta*)—Is the famous variable of this constellation, and one of the most remarkable, but not visible from our latitude; here we give the table of the variations observed, the years, and the names of the observers:

YEARS.	MAGNITUDE.	OBSERVERS.	YEARS.	MAGNITUDE.	OBSERVERS.
1677	4	Halley.	1858	2.6	Powell.
1751	2	Lacaille.	1860	3.5	Tebbutt.
1811-15	4	Burchell.	1862	4.3	Tebbutt.
1822-26	2	Fallows, Brisbane.	1864	5.0	Tebbutt.
1827	1	Burchell.	1866	5.6	Tebbutt.
1828-33	2	Johnson, Taylor.	1868	6.1	Tebbutt.
1834-37	1½	John Herschel.	1870	6.5	Tebbutt, Gould.
1838-42	1	Maclear.	1872	6.8	Tebbutt.
1842	Nearly as Sirius	Maclear.	1874	7.0	Tebbutt.
1844-54	1	Jacob, Gilliss.	1876	7.2	Gould.
1856	1.5	Powell.	1878	7.4	Gould.

Since 1867 it is not visible to the naked eye. Is the variation periodical? Will it come brighter again? So far we do not know.

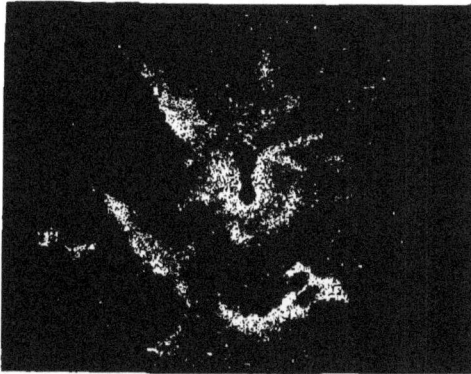

Fig. 136.—The Great Nebula in Argo Navis.

η (*Eta*) is near the center of the famous nebula 2197 of J. Herschel's general catalogue, which occupied nearly one degree; like the great nebula of Orion, it can not be resolved into stars. Sir J. Herschel undertook and succeeded in illustrating this beautiful object; it shows 1,203 stars, all measured and properly located by this *habile* observer; the light must take thousands of years to come from there, and is most likely not as it appears to us now. The region of the "Milky Way" near this nebula is one of the richest in stars, and J. Herschel counted 250 of them in a field of 15 minutes in diameter and 147,000 in 47 square degrees! (Fig. 136.)

CENTAURUS.

Centaurus, or the Centaur, is most likely the same Cheiron already spoken of in Sagittarius. The Centaurs were a nomadic tribe of Thessalian race; as they were excellent horsemen, they were fabled as being half man and half horse.

This constellation appears in the Greek sphere in Eudoxus' time, and is certainly older.

DESIGNA-TION.	MAGNI-TUDE.	POSITION R. A. 1880 h. m.	DECL. ° '	DESIGNA-TION.	MAGNI-TUDE.	POSITION R. A. 1880 h. m.	DECL. ° '
α dbl.	1.0 yel.	14.31	−60.20	ξ2	4.8	13. 0	−49.16
β	1.5	13.55	59.48	o1	5.2	11.26	58.47
γ dbl.	2.5	12.35	48.18	o2	5.5	11.26	58.51
δ	2.8	12. 2	50. 3	π	4.3	11.16	53.50
ε	2.6	13.32	52.51	ρ	4.5	12. 5	51.42
ζ dbl.	2.7	13.48	46.42	σ	4.3	12.22	49.34
η	2.5	14.28	41.38	τ	4.4	12.31	47.53
θ	2.3	14. 0	35.47	ν1	4.2	13.51	44.13
ι	3.0	13.14	36. 5	ν2	5.0	13.54	45. 1
κ	3.3	14.51	41.37	φ	4.1	13.51	41.31
λ	3.4	11.30	62.21	χ	4.8	13.59	40.36
μ	3.4	13.42	41.53	ψ	4.4	14.13	37.20
ν	3.7	13.42	41. 5	ω	cl,	13.20	46.51
ξ1	5.8	12.57	48.53	R	6.V. red	14. 8	59.20

NOTES.

This constellation is too far south to be seen entirely from the United States; it contains the nearest star to us, which is also one of the most interesting doubles.

α (*Alpha*)—Binary; magnitudes 1st and 2d; the components revolve around each other in 84 years; the plane of the revolution forms an angle of 79° with the plane perpendicular to our vision, and consequently the ellipse appears more than twice as long as it really is. We give herewith the principal measures of the angle and distances of the two stars since John Herschel's observations:

YEAR.	OBSERVERS.	ANGLE.	DISTANCE.
1833	J. Herschel........................	217°5	18".7
1840	Maclear.........	223°2	14".7
1850	Jacob............................	250°7	6".0
1856	Jacob............................	307°0	3".9
1860	Powell...........................	347°0	5".6
1870	Powell...........................	22°0	10".3
1878	Gill.............................	100°0	1".9
1880	Gruls	165°0	3".1

The companion was noted of 4th magnitude before 1830 by Feuillee, Lacaille, Brisbane and Dunlop; of 3d magnitude by J. Herschel, in 1835; it is now of the 2d magnitude; its color is orange. α (*Alpha*) is also in rapid motion, 3".67 per year; 6 minutes in 100 years. If this motion continues it will pass very near β (*beta*); in 12,000 years it will be in the Southern Cross, and in the 500th century it will be near γ (*gamma*) of Argo Navis

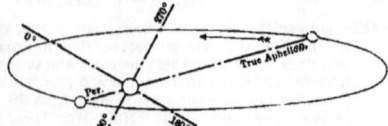

Fig. 137.—Apparent Orbit of α

We give below the measures of its parallax at different times:

OBSERVERS.	PARALLAX.	OBSERVERS.	PARALLAX.
Henderson, 1838................	1″.16 ±0″.11	Gill, 1883....................	0″.747 ±0″.013
Henderson and Maclear, 1842....	0″.913 ±0″.064	Gill, 1883....................	0″.765 ±0″.017
Maclear, 1851.............	....0″.919 ±0″.034	Elkin, 1882.............	0″.783 ±0″.028
Mœsta, 1864................	0″.521 ±0″.066	Elkin, 1883................	0″.676 ±0″.027

The last four are the result of careful micrometrical measures, and in taking their average, 0″.75, it represents 275,000 times the distance from the earth to the sun, or 25 trillions of miles, and the light travels 4 years and 128 days to reach us (Revue d'Ast., 1889; page 443). A fast train going at the rate of 60 miles an hour would have to run for more than 47 million 500 thousand years without stopping before arriving on this sun, which, as we said before, is the nearest to us. a (*Alpha*) is the standard of stars of 1st magnitude. From the photometrical measures of Sir J. Herschel, the light of a (*alpha*) equals 1-27,000th of the light of the full moon; the full moon 1-800,000th of the light of the sun, consequently it would

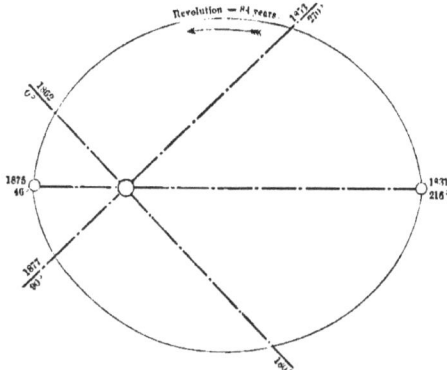

Fig. 138.—Real Orbit of a

take 22 millions of stars of the same magnitude to produce as much light as the sun; it would take only 5,400 stars like Sirius to give the same result. Seen from a Centauri, our sun is only a star of the 2d magnitude; if they were revolving around their center of gravity the revolution would take about 13 millions of years to be accomplished.

β (*Beta*)—The parallax of this star is given as below:

Maclear, in 1842-3...... 0″.470 ±0″.444
Mœsta, in 1860-4 .... 0″.173 ±0″.070
Gill, in 1882 ....−0″.017 ±0″.072

The last one considered as most correct, being "negative," indicates that β (*beta*) is too far to calculate its distance (Revue d'Ast., 1889; page 449).

ω (*Omega*)—Is the beautiful cluster, visible to the naked eye as a star of the 4th magnitude; it is the *finest cluster of the heavens*; its shape is nearly spherical and it contains several thousand stars in a size about two-thirds of the apparent disc of the moon. It could be seen near the horizon passing the meridian at midnight, the 10th of April.

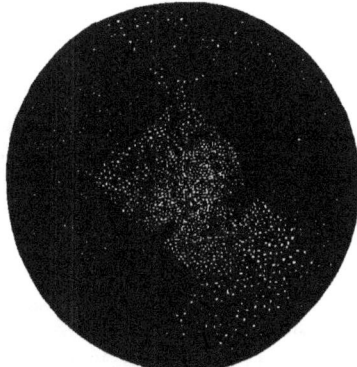

Fig. 139.—Cluster ω

LUPUS.

Lupus, or the Wolf, is an old constellation, mention of it being made by Eudoxus; the Greeks also called it the Ferocious Beast. It most likely represents Lycaon, King of Arcadia, struck by lightning and changed into a wolf by Jupiter, whom he had offended.

DESIGNA-TION.	MAGNI-TUDE.	POSITION		DESIGNA-TION.	MAGNI-TUDE.	POSITION	
		R. A. 1880 h. m.	DECL. o '			R. A. 1880 h. m.	DECL. o '
α	2.6	14.34	—46.52	λ	4.8	15. 1	—44.48
β	2.8	14.51	42.39	μ dbl.	4.8	15.10	47.26
γ dbl.	3.2	15.27	40.46	π dbl.	4.3	14.57	46.35
δ	3.7 red	15.13	40.13	ρ	4.5	14.30	48.54
ε dbl.	3.7	15.14	44.15	φ1	3.6 red	15.14	35.50
ζ dbl.	3.6	15. 4	51.38	φ2	5.1	15.15	36.26
η dbl.	3.7	15.52	38. 3	χ	4.2	15.43	33.15
θ	4.9	15.59	36.29	4 Lac.	5.V.	15.35	34.19
ι	3.8	14.12	45.30	6380 Lac.	5.9 red	15.21	46.19
κ dbl.	4.2	15. 4	48.17				

NOTES.

This constellation is almost entirely visible from New York, Chicago and San Francisco, but very close to the horizon; it is composed of stars of small magnitude; the brightest one, α (alpha), being only a little above the 3d.

φ1 (Phi)—Is very red; it was noted of the 4th magnitude in 1751; of the 5th in 1825; of the 4.2 in 1836; of 4½ds in 1860, and of 3.6 in 1878.

ARA.

Ara, or the Altar, appeared in the sphere of Eudoxus, and was introduced most likely to perpetuate a sacrifice to the gods—perhaps apropos of the Argonauts' expedition.

DESIGNA-TION.	MAGNI-TUDE.	POSITION		DESIGNA-TION.	MAGNI-TUDE.	POSITION	
		R. A. 1880 h. m.	DECL. o '			R. A. 1880 h. m.	DECL. o '
α dbl.	2.9	17.22	—49.47	ε2	5.9	16.53	—53. 3
β	2.8	17.15	55.25	ζ	3.2 red	16.49	55.48
γ dbl.	3.6	17.15	56.15	η	3.8	16.39	58.49
δ dbl.	3.7	17.20	60.35	θ dbl.	3.9	17.57	50. 6
ε1	4.2	16.50	52.58	*	5.V.	17.30	45.24

NOTE.

This constellation is too far south to be seen entirely on our planisphere, α (alpha) and θ (theta) only touching the horizon at 9 p. m., at the end of July and beginning of August.

CORONA AUSTRALIS.

Corona Australis, or the Southern Crown, is one of the forty-eight constellations of the ancients, being already mentioned by Eudoxus and Aratus.

DESIGNA-TION.	MAGNI-TUDE.	POSITION		DESIGNA-TION.	MAGNI-TUDE.	POSITION	
		R. A. 1880 h. m.	DECL. o '			R. A. 1880 h. m.	DECL. o '
α	4.2	19. 1	—38. 5	γ bin.	4.6	18.58	37.14
β	4.1	19. 2	39.32				

NOTES.

This constellation contains only three stars above the 5th magnitude.

γ (Gamma)—Binary; magnitudes 5.5 and 5.5; distance, 1".5; close pair; in very rapid orbital motion; revolution, 55 years.

INDUS.

Indus, or the Indian, is a modern constellation, formed by Bayer in 1603.

DESIGNA-TION.	MAGNI-TUDE.	POSITION		DESIGNA-TION.	MAGNI-TUDE.	POSITION	
		R. A. 1880	DECL.			R. A. 1880	DECL.
		h. m.	° '			h. m.	° '
α dbl.	3.1	20.29	−47.43	ε	5.2	21.54	−57.17
β	3.7	20.45	58.54	ζ	5.3	20.41	46.40
γ	6.3	21.18	55.11	η	4.7	20.35	52.21
δ	4.8	21.50	55.34	θ dbl.	4.6	21.11	53.57

NOTES.

With the exception of α (alpha) and ζ (zeta) all the stars of this constellation are too far south to be seen from our latitude.

γ (Gamma)—Was of the 4½th magnitude, but is now hardly visible to the naked eye.

ε (Epsilon)—Is in rapid motion, 4″.60 per year, and will reach the south pole in 50,000 years if nothing occurs to retard its present motion. Its parallax has been tried and found to be:

By Elkin and Gill, in 1883...0″.274±0″.027
By Elkin, in 1883...0″.124±9″.019

The average, 0″.22, represents 937,000 times the distance of the earth from the sun, or 87 trillions 500 billions of miles, and it takes the light 14 years and 146 days to reach us.

GRUS.

Grus, or the Goose, is a modern constellation, formed by Bayer in 1603.

DESIGNA-TION.	MAGNI-TUDE.	POSITION		DESIGNA-TION.	MAGNI-TUDE.	POSITION	
		R. m. 1880	DECL.			R. A. 1880	DECL.
		h. m.	° '			h. m.	° '
α	2.0	22. 1	−47.32	ε	3.5	22.41	−51.57
β	2.3 red	22.35	47.31	ζ	4.0	22.54	53.24
γ	3.0	21.47	37.56	η	5.1	22.38	54. 8
δ1	4.2	22.22	44. 7	θ dbl.	4.2	23. 0	44.10
δ2 dbl.	4.4	22.23	44.22	ι	3.9	23. 4	45.54

NOTE.

The southern part of this constellation is not visible from our latitude.

PHŒNIX.

Phœnix, or the Phenix, is a modern constellation, formed by Bayer in 1603.

DESIGNA-TION.	MAGNI-TUDE.	POSITION		DESIGNA-TION.	MAGNI-TUDE.	POSITION	
		R. A. 1880	DECL.			R. A. 1880	DECL.
		h. m.	° '			h. m.	° '
α	2.4	0.20	−42.58	ε	3.8	0. 3	−46.24
β	3.3	1. 1	47.22	ζ dbl.	4.2	1. 3	55.53
γ	3.4 red	1.23	43.56	η dbl.	4.5	0.38	58. 7
δ	4.0	1.26	49.42				

NOTE.

The southern part of this constellation is not visible from our latitude.

HOROLOGIUM.

Horologium, the Horoscope or the Clock, is a new constellation, formed by Lacaille in 1752.

DESIGNA-TION.	MAGNI-TUDE.	POSITION		DESIGNA-TION.	MAGNI-TUDE.	POSITION	
		R. A. 1880	DECL.			R. A. 1880	DECL.
		h. m.	° '			h. m.	° '
α	3.8	4.10	−42.35	γ	6.1	2.43	−64.13
....	5.2	2.57	64.33	δ	5.3	4. 7	42.18
β { 1320 Lac.	7.0	3.56	44.15				

NOTE.

Only a part of this constellation rises above the horizon in our latitude.

RETICULUM.

Reticulum, or the Reticule, is a new constellation, formed by Lacaille in 1752.

DESIGNA-TION.	MAGNI-TUDE.	POSITION R. A. 1880 h. m.	DECL. ° '	DESIGNA-TION.	MAGNI-TUDE.	POSITION R. A. 1880 h. m.	DECL. ° '
α dbl.	3.3	4.13	−62.46	ε	4.6	4.15	−59.35
β	3.9	3.43	65.11	ζ¹	5.9	3.15	63. 2
γ	4.7 red	3.59	62.30	ζ²	5.7	3.16	62.58
δ	4.7	3.57	61.44				

NOTES.

This southern constellation is never visible from our latitude.

ζ (*Zeta*)—Is a double star visible to the naked eye; magnitudes 5.7 and 5.9; in rapid motion toward the region where we come from.

XIPHIAS OR DORADO.

Xiphias, Dorado, or the Craver, is a constellation formed by Bayer in 1603.

DESIGNA-TION.	MAGNI-TUDE.	POSITION R. A. 1880 h. m.	DECL. ° '	DESIGNA-TION.	MAGNI-TUDE.	POSITION R. A. 1880 h. m.	DECL. ° '
α dbl.	3.1	4.32	−55.17	δ	4.5	5.45	−65.46
β	3.9	5.33	62.34	ε	5.1	5.50	66.56
γ	4.4	4.13	51.47	ζ	4.8	5. 4	57.38

NOTE.

No star of this constellation ever rises above the horizon in our latitude.

PLUTEUM PICTORIS.

Pluteum Pictoris, or the Easel, is a new constellation, formed by Lacaille in 1752.

DESIGNA-TION.	MAGNI-TUDE.	POSITION R. A. 1880 h. m.	DECL. ° '	DESIGNA-TION.	MAGNI-TUDE.	POSITION R. A. 1880 h. m.	DECL. ° '
α	3.5	6.47	−61.49	γ	4.7	5.48	−56.12
β	3.9	5.45	51. 7	δ	5.2	6. 8	54.57

NOTE.

This southern constellation is never visible from our latitude.

PISCIS VOLANS.

Piscis Volans, or the Flying Fish, first appears in Bayer's atlas in 1603.

DESIGNA-TION.	MAGNI-TUDE.	POSITION R. A. 1880 h. m.	DECL. ° '	DESIGNA-TION.	MAGNI-TUDE.	POSITION R. A. 1880 h. m.	DECL. ° '
α	4.2	9. 1	−65.55	ε dbl.	4.5	8. 8	−68.16
β	3.9	8.25	65.44	ζ dbl.	4.3	7.44	72.19
γ dbl.	3.8	7.10	70.18	η dbl.	4.5	8.22	73. 0
δ	4.1	7.17	67.44	θ dbl.	4.5	8.39	70. 0

NOTES.

This constellation is never visible from our latitude.

In this constellation α (*alpha*) varies from the 3¼th to the 5th magnitude; β (*beta*), γ (*gamma*) and δ (*delta*) from the 4th to the 5th magnitude; ε (*epsilon*) from the 4.5 to the 6th magnitude, and ζ (*zeta*) from the 4.3 to the 5.7 magnitude.

CHAMÆLEON.

Chamæleon, or the Chameleon, is a constellation formed by Bayer in 1603.

DESIGNA-TION.	MAGNI-TUDE.	POSITION		DESIGNA-TION.	MAGNI-TUDE.	POSITION	
		R. A. 1880 h. m.	DECL. ° ′			R. A. 1880 h. m.	DECL. ° ′
α	4.2	8.22	−76.32	γ	4.4	10.34	−77.59
β	4.6	12.11	78.39	δ dbl.	4.8	10.45	79.54

NOTE.

This constellation is never visible from our latitude.

APIS or MUSCA.

Apis, the Bee (also called Musca, the Fly), is a small constellation formed by Bayer in 1603.

DESIGNA-TION.	MAGNI-TUDE.	POSITION		DESIGNA-TION.	MAGNI-TUDE.	POSITION	
		R. A. 1880 h. m.	DECL. ° ′			R. A. 1880 h. m.	DECL. ° ′
α	2.9	12.30	−68.29	ζ¹	6.5 red	12.15	−67.38
β	3.4	12.39	67.27	ζ²	5.8	12.15	66.51
γ	4.0	12.25	71.28	λ	3.8	11.40	66. 4
δ	3.7	12.54	70.54	μ	5.3 red	11.43	66. 9
ε	4.7	12.11	67.17				

NOTE.

This constellation is never visible from our latitude.

CRUX.

Crux, or the Cross, was part of the constellation Centaurus; it is mentioned by Hipparchus in the 1st century B. C., and was introduced by Augustin Royer in 1679, but a cross already appeared in Bayer's Atlas in 1603.

DESIGNA-TION.	MAGNI-TUDE.	POSITION		DESIGNA-TION.	MAGNI-TUDE.	POSITION	
		R. A. 1880 h. m.	DECL. ° ′			R. A. 1880 h. m.	DECL. ° ′
α dbl.	1.6	12.20	−62.26	η dbl.	4.7	12. 1	−63.56
β	1.8	12.41	59. 2	θ1	4.7	11.57	62.39
γ dbl.	2.0 red	12.25	56.26	θ2	5.3	11.58	62.30
δ	3.4	12. 9	58. 5	ι	5.7	12.39	60.19
ε	4.0	12.15	59.44	κ (cl.)	6.7 red	12.47	59.43
ζ dbl.	4.6	12.12	63.20	*	8.5 red	12.40	59. 2

NOTES.

This beautiful Southern Constellation, traversed by the "Milky Way," is very rich and contains no less than three stars of the 1st magnitude, but it is too far south to be seen from our latitude.

κ (Kappa)—Is hardly visible to the naked eye; offers in the field a beautiful cluster of 110 stars of every color, among them one ruby, one marine blue, two emerald and three pale green; the white stars shine like diamonds.

γ (Gamma)—Has a beautiful spectrum, very interesting, similar to the spectrum of α (alpha) of Hercules and Betelgeuse, and some lines indicating the presence of water in its atmosphere in a vaporous state. The dark spot in the "Milky Way" between the Cross and Apis is the famous "Coal Sack," also called Macula Magellanæ.

CIRCINUS.

Circinus, the Compasses, is a constellation formed by Lacaille in 1752.

DESIGNA-TION.	MAGNI-TUDE.	POSITION		DESIGNA-TION.	MAGNI-TUDE.	POSITION	
		R. A. 1880 h. m.	DECL. ° ′			R. A. 1880 h. m.	DECL. ° ′
α dbl.	3.5	14.33	−64.27	γ dbl.	5.2	15.14	−58.53
β	4.7	15. 8	58.21				

NOTE.

This constellation is never visible from our latitude.

TRIANGULUM AUSTRALIS.

Triangulum Australis, or the Southern Triangle, is a constellation formed by Bayer in 1603.

| DESIGNA-TION. | MAGNI-TUDE. | POSITION | | DESIGNA-TION. | MAGNI-TUDE. | POSITION | |
		R. A. 1880 DECL. h. m. ° '				R. A. 1880 DECL. h. m. ° '	
α	2.2 red	16.36	—68.48	γ	3.1	15. 8	—68.14
β	3.1	15.44	63. 3				

NOTE.

This constellation is never visible from our latitude.

APUS.

Apus, the Indian Bird or Paradise Bird, is a constellation formed by Bayer in 1603.

| DESIGNA-TION. | MAGNI-TUDE. | POSITION | | DESIGNA-TION. | MAGNI-TUDE. | POSITION | |
		R. A. 1880 DECL. h. m. ° '				R. A. 1880 DECL. h. m. ° '	
α	4.0	14.33	—78.32	γ	3.9	16.15	—78.37
β dbl.	4.5 red	16.26	77.16				

NOTE.

This constellation is never visible from our latitude.

PAVO.

Pavo, or the Peacock, is of recent origin, and was formed into a constellation by John Bayer in 1603.

| DESIGNA-TION. | MAGNI-TUDE. | POSITION | | DESIGNA-TION. | MAGNI-TUDE. | POSITION | |
		R. A. 1880 DECL. h. m. ° '				R. A. 1880 DECL. h. m. ° '	
α	2.1	20.16	—57. 7	ι	5.8	17.59	—62. 1
β	3.3	20.34	66.38	κ	4.V.	18.44	67.23
γ	4.5	21.16	65.55	λ	4.3	18.41	62.20
δ	3.5 red	19.57	66.29	μ¹	5.9 red	19.48	67.16
ε	4.0	19.46	73.15	μ²	5.6 red	19.50	67.17
ζ dbl.	4.2	18.29	71.32	ν	4.8	18.20	62.21
η	3.8	17.34	64.40	π	4.6	17.57	63.40
θ	6.1	18.37	65.12	*	7.0 very red	17.33	57.40

NOTES.

This constellation is never visible from our latitude.

α (*Alpha*)—Is a bright star of nearly the 1st magnitude.

ι (*Iota*)—Does not appear in the last catalogue of Lacaille, nor in Brisbane; it must have changed magnitude; it is now of the 6th.

κ (*Kappa*)—Varies from the 4th to the 6th magnitude.

TOUCAN.

The Toucan was formed into a constellation by John Bayer in 1603.

| DESIGNA-TION. | MAGNI-TUDE. | POSITION | | DESIGNA-TION. | MAGNI-TUDE. | POSITION | |
		R. A. 1880 DECL. h. m. ° '				R. A. 1880 DECL. h. m. ° '	
α	2.8 red	22.10	—60.51	ε	4.3	23.54	—66.14
β dbl.	3.7	0.26	63.37	ζ	4.1	0.14	65.35
γ	4.0	23.10	58.54	52 Herschel cl.		0.19	72.45
δ dbl.	4.8	22.19	65.34				

NOTES.

ζ (Zeta)—Mr. Elkin, of Yale College, obtained a parallax of this star = 0″.057 ± 0″.019; this parallax is considered too uncertain to determine its distance.

This constellation is never visible from our latitude. It has a beautiful cluster visible to the naked eye; a bright double star is near the center. This cluster (Fig. 140), the 52d of John Herschel's catalogue, is also called *Hersch.* (See our map of the Southern Constellations, page 66.)

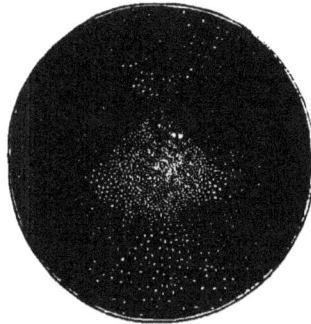

Fig. 140.—Cluster 52 Herschel.

HYDRUS.

Hydrus, the male Hydra or the Snake, is one of the constellations of Bayer's atlas (1603).

DESIGNA- TION.	MAGNI- TUDE.	POSITION		DESIGNA- TION.	MAGNI- TUDE.	POSITION	
		R. A. 1880 h. m.	DECL. ° ′			R. A. 1880 h. m.	DECL. ° ′
α	2.9	1.55	−62. 9	δ	4.1 red	2.20	−69.12
β	2.7	0.19	77.56	ε	4.2	2.38	68.47
γ	3.2 red	3.49	74.36				

NOTE.

This constellation is never visible from our latitude. It contains the "Nebecula Minor," which occupies nearly 10 square degrees, in which J. Herschel discovered 32 nebulæ, 6 clusters, and 200 distinct stars.

MONS MENSÆ.

Mons Mensæ, or Mount of the Table, is a constellation formed by Lacaille in 1752.

DESIGNA- TION.	MAGNI- TUDE.	POSITION		DESIGNA- TION.	MAGNI- TUDE.	POSITION	
		R. A. 1880 h. m.	DECL. ° ′			R. A. 1880 h. m.	DECL. ° ′
α	5.3	6.14	− 74.43	γ (lbl.)	5.6	5.37	−76.26
β	5.7	5. 5	71.29	δ	5.8	4.27	80.30

NOTES.

This constellation is never visible from our latitude.

Between Mens Mensæ and Hydrus is the "Nebecula Major," an immense cluster 200 times larger than the apparent disc of the moon, and covering no less than 42 square degrees. J. Herschel found in it 284 nebulæ, 64 clusters, and 582 separate stars.

The Nubeculæ Major and Minor are also called *Magellanic Clouds.*

OCTANS.

Octans, the Octant, is the constellation occupying the South Pole, and had been formed by Lacaille in 1752.

DESIGNA- TION.	MAGNI- TUDE.	POSITION		DESIGNA- TION.	MAGNI- TUDE.	POSITION	
		R. A. 1880 h. m.	DECL. ° ′			R. A. 1880 h. m.	DECL. ° ′
α	5.6	20.50	−77.28	ν	3.8	21.28	−77.54
β	4.4	22.33	82. 1	σ	5.8	18.16	89.17
γ	5.5	23.45	82.41	τ	6.0	23. 9	88. 8
δ	4.7	14. 7	83. 7				

NOTE.

This last constellation, occupying the south pole, and consequently invisible from our latitude, contains no star of any importance; the brightest is no longer α (alpha) but β (beta).

OLD AND NEW CONSTELLATIONS IN CHRONOLOGICAL ORDER.

CONSTELLATIONS.	WHEN FIRST CITED.
Ursa Major	Job XXXVIII, 31 (17th century B. C.?); Homer (9th century B. C.).
Orion	Job IX, 9; Homer; Hesiod.
Pleiades	Job XXXVIII, 31; Homer; Hesiod.
Hyades	" " "
Sirius and Canis Major	Homer; Hesiod.
Aldebaran and Taurus	" "
Arcturus and Bootes	" "
Ursa Minor	Thales (7th century B. C.); Eudoxus (4th century B. C.); Aratus (3d century B. C.).
Draco	Eudoxus; Aratus.
Engonasi (called Hercules afterward)	"
*Cerberus and Ramus	"
Corona Borealis	"
Ophiuchus and Serpens	"
Scorpio	"
Spica and Virgo	"
Gemini	"
Procyon	"
Leo	"
Capella and Auriga	"
Cepheus	"
Cassiopea	"
Andromeda	"
Pegasus	"
Deltoton or Triangulum	"
Pisces	"
Perseus	"
Lyra	"
Cygnus	"
Aquila	"
Aquarius	"
Capricornus	"
Sagittarius	"
Sagitta	"
Delphinus	"
Lepus	"
Argo Navis	"
Canobus (called later Canopus)	"
Eridanus	"
Cetus	"
Piscis Notius	"
Corona Australis	"
Ara	"
Centaurus	"
Lupus	"
Hydra	"
Crater	"
Corvus	"

* These constellations, not being recognized by the B. A. Catalogue, are not described in this Handbook; the stars by which they were formed are inserted in the constellations from which they were taken.

CONSTELLATIONS.	WHEN FIRST CITED.
Libra.....................Manetho (3d century B. C.); Geminus (1st century B. C.).
Coma Berenices...........Callimachus (3d century B. C.); Erasthosthenes (3d century B. C.); Conon.
Crux.......................Hipparchus, 130 years B. C.
Propus (1 Geminorum).... "
Equuelus................. "
Præsepe................. "
Caput Medusæ........... "
Antinous...................During the reign of Emperor Adrian, 132 years A. D.
Hercules..................Hyginus, 1485, A. D.
Pavo.......................John Bayer, 1603, A. D.
Toucan................... "
Grus...................... "
Phœnix.... .... "
Xiphias or Dorado........ "
Pisces Volans. .... "
Hydrus................... "
Chamæleon............... "
Apis...................... "
Apus...................... "
Triangulum Australis...... "
Indus..................... "
Columba. .... "
Crux .... "
*Musca. .. ..Bartschius, 1624, A. D.
Camelopardalus........... "
Monoceros .... " already mentioned in 1564.
Canes Venatici... ....Hevelius, 1660, A. D.
Vulpecula................. "
Leo Minor........... . .... "
Lynx .... "
Robur Caroli II .. ..Halley, 1677, A. D.
Cor Caroli II.............. "
Crux................. .....Augustin Royer, 1679, A. D., already mentioned by the ancients, and drawn in Bayer's Atlas, 1603, A. D.
Columba. ....Generally attributed to Augustin Royer, already drawn in Bayer's Atlas.
*Sceptrum Brandenburgicum ....Godfried Kirch, 1688, A. D.
Nubecula Major... ....Hevelius, 1690. A. D.
Nubecula Minor........... "
Camelopardalus. .... " already mentioned by Bartschius.
Lacerta.... .... "
Sextans Uraniæ........... "
Scutum Sobiesii .... "
*Triangulum Minus ...... "
*Mons Menalus............Flamsteed, 1725, A. D.
Apparatus Sculptoris......Lacaille, 1752.
Apparatus Chemicus...... "
Horologium............... "
Reticulum .... "
*Cæla Scalptoris.......... "
Pluteum Pictoris .... "
*Pixis Nautica............. "
Antlia Pneumatica........ "

* These constellations, not being recognized by the B. A. Catalogue, are not described in this Handbook; the stars by which they were formed are inserted in the constellations from which they were taken.

CONSTELLATIONS.	WHEN FIRST CITED.
Octans	Lacaille, 1752.
Circinus	"
*Norma Regula	"
*Tubus Astronomicus	"
*Microscopium	"
Mons Mensæ	"
*Tartus Solitarius	Lalande, 1774, A. D.
*Messium	"
*Rangifer	Lemonnier, 1776, A. D.
*Taurus Poniatowii	Poczobut, 1777, A. D.
*Telescopium Herschelii	Hell, 1789, A. D.
*Harpa Georgii	"
*Quadrans Muralis	Lalande, 1795.
*Honores Freiderici	Bode, 1798, A. D.
*Machina Electrica	"
*Officina Typographica	"
*Globus Aerostatis	Lalande, 1798.
*Felis	" 1799.

* These constellations, not being recognized by the B. A. Catalogue, are not described in this Handbook; the stars by which they were formed are inserted in the constellations from which they were taken.

NAMES GIVEN TO THE PRINCIPAL STARS.

Achernar, α (alpha) Eridani.
Adara, ε (epsilon) Canis Majoris.
Albireo, β (beta) Cygni.
Alchiba, α (alpha) Corvi.
Alcyone, η (eta) Tauri, Pleiad.
Aldebaran, α (alpha) Tauri.
Alderamin, α (alpha) Cephei.
Al-fard or Alphard, α (alpha) Hydræ.
Algeiba, γ (gamma) Leonis.
Algenib, γ (gamma) Pegasi.
Algol, β (beta) Persei.
Algores, δ (delta) Corvi.
Alhena, γ (gamma) Geminorum.
Alioth, ε (epsilon) Ursæ Majoris.
Alkaid, or Benetnash, η (eta) Ursæ Majoris.
Alkes, α (alpha) Crateris.
Almach, γ (gamma) Andromedæ.
Alniban, ε (epsilon) Orionis.
Alphecca, α (alpha) Coronæ Borealis.
Alpherat, α (alpha) Andromedæ.
Alphirk, β (beta) Cephei.
Alshain, β (beta) Aquilæ.
Altair, α (alpha) Aquilæ.
Alwaid, β (beta) Draconis.
Antares, α (alpha) Scorpionis.
Baten Kaitos, ζ (zeta) Ceti.
Bellatrix, γ (gamma) Orionis.
Benetnash, see Alkaid.
Betelgeuse, α (alpha) Orionis.
Canopus, α (alpha) Argo Navis.
Capella, α (alpha) Aurigæ.
Castor, α (alpha) Geminorum.
Cebalrai, β (beta) Ophiuchi.
Chaph, β (beta) Cassiopeiæ.
Cor Caroli II, α (alpha) Canum Venatici.
Cor Hydræ, see Al-fard.
Cor Leonis, see Regulus.
Cor Scorpionis, see Antares.
Cursa, β (beta) Eridani.
Deneb, α (alpha) Cygni.
Denebola, β (beta) Leonis.
Diphda, β (beta) Ceti.
Dog Star, The (see Sirius).
Dubhe, α (alpha) Ursæ Majoris.
Enif, ε (epsilon) Pegasi.
Errai, γ (gamma) Cephei.
Etanin, γ (gamma) Draconis.
Fomalhaut, α (alpha) Piscis Australis.
Gemma, the Jewel see Alphecca.
Gomeisa, β (beta) Canis Minoris.
Hamal, α (alpha) Arietis.
Homan, ζ (zeta) Pegasi.
Hyades, in Taurus.
Izar, ε (epsilon) Boötis.
Jewel, The (see Alphecca).
Kaus Australis, ε (epsilon) Sagittarii.

Kiffa Australis, α (alpha) Libræ.
Kiffa Borealis, β (beta) Libræ.
Kochab, β (beta) Ursæ Minoris.
Korneforos, β (beta) Herculis.
Markab, α (alpha) Pegasi.
Mebsuta, ε (epsilon) Geminorum.
Megrez, δ (delta) Ursæ Majoris.
Menkalinan, β (beta) Aurigæ.
Menkar, α (alpha) Ceti.
Merak, β (beta) Ursæ Majoris.
Mesartim, γ (gamma) Arietis.
Mintaka, δ (delta) Orionis.
Mira, o (omicron) Ceti.
Mirach, β (beta) Andromedæ.
Mirfak, α (alpha) Persei.
Mirzam, β (beta) Canis Majoris.
Mizar, ζ (zeta) Ursæ Majoris.
Muphrid, η (eta) Boötis.
Nath, β (beta) Tauri.
Nekkar, β (beta) Boötis.
Phact, α (alpha) Columbæ.
Phegda, γ (gamma) Ursæ Majoris.
Pleiades: Alcyone, Electra, Maia, Merope, Taygete, Atlas, Pleione, Celæno, Asterope.
Propus, 1 Geminorum.
Ras-Alhague, α (alpha) Ophiuchi.
Ras-Algethi, α (alpha) Herculis.
Regulus, α (alpha) Leonis.
Rigel, β (beta) Orionis.
Rotanev, β (beta) Delphini.
Sadalmelik, α (alpha) Aquarii.
Sadalsund, β (beta) Aquarii.
Salaphat, γ (gamma) Lyræ.
Saidak. (See Alcor).
Scheat, β (beta) Pegasi.
Schedar, α (alpha) Cassiopeiæ.
Secunda Giedi, α2 (alpha2) Capricorni.
Sheilak, β (beta) Lyræ.
Sheratan, β (beta) Arietis.
Sirius, α (alpha) Canis Majoris.
Skat, δ (delta) Aquarii.
Spica, α (alpha) Virginis.
Sualocin, α (alpha) Delphini.
Talitha, ι (iota) Ursæ Majoris.
Tarazed, γ (gamma) Aquilæ.
Thuban, α (alpha) Draconis.
Vega, α (alpha) Lyræ.
Vindemiatrix, ε (epsilon) Virginis.
Wesat, δ (delta) Geminorum.
Zaurac, γ (gamma) Eridani.
Zavijava, β (beta) Virginis.
Zosma, δ (delta) Leonis.
Zuben el Chameli, (see Kiffa Borealis).
Zuben el Genubi, (see Kiffa Australis).
Zuben Hakrakl, γ (gamma) Libræ.

THE PRINCIPAL BINARY STARS.

DESIGNATION.	MAGNI-TUDES.	PERIODS OF REVOLUTION.	COLORS.
κ (*Kappa*) Pegasi	4-9	11 years.....	White and purple.
δ (*Delta*) Equulei	4.5-5	12 "	Both white.
Σ 3130 Lyræ	7.4-11	16 "	Both white.
42 Comæ Berenicis	6-6	25 "	Both white.
8 Sextans	5.6-6.5	33 "	Both white.
ζ (*Zeta*) Herculis	3-6	34½ "	Yellow and orange.
Σ 3121 Cancri	7.2-7.5	39 "	White and yellow.
η (*Eta*) Coronæ Borealis	5.3-5.5	41 "	White and gold yellow.
Σ 2173 Ophiuchi	6-6	45 "	Both yellow.
Sirius	1-9	49(?)"	Both white.
OΣ 527	7-8	54 "	Bluish and white.
γ (*Gamma*) Coronæ Australis	5.5-5.5	55 "	Gold-yellow.
ζ (*Zeta*) Cancri (trinary) { A B { A C	5.5-6.2 5.5-6.6	60 " ..... 600 "	Both yellow. Both yellow.
ξ (*Zi*) Ursæ Majoris	4-5	60 "	Yellow and ash color.
OΣ 234 Ursæ Majoris	7-7.8	68 "	Both white.
OΣ 298 Boötis	7-7.4	69 "	Both white.
α (*Alpha*) Centauri	1-2	84(?)"	White and yellow.
70 Ophiuchi	4.5-6	90 "	Yellow and rose.
OΣ 235 Ursæ Majoris	6-7	94 "	Both white.
γ (*Gamma*) Coronæ Borealis	4-7	95 "	Yellow and purple.
ξ (*Zi*) Scorpionis (trinary) A B	5-5.2	96 "	Both yellow.
Σ 2107 Herculis	6.5-8.5	98 "	Yellow and blue.
Σ 3062 Cassiopeiæ	6.5-7.5	104 "	Yellow and olive.
φ (*Phi*) Ursæ Majoris	5-5.5	115 "	Both yellow.
ω (*Omega*) Leonis	6-7	124 "	White and blue.
25 Canum Venatici	6-7	124 "	White and blue.
ξ (*Zi*) Boötis	4.5-6.5	127 "	Yellow and red.
4 Aquarii	6-7	130 "	Both yellow.
o² (*Omicron²*) Eridani (trinary) B C	9.5-10.5	139 "	Both yellow.
η (*Eta*) Cassiopeiæ	4.2-7	167 "	Yellow and purple.
γ (*Gamma*) Virginis	3-3	175 "	Both yellow.
τ (*Tau*) Ophiuchi	5.2-6	218 "	Both white.
44 ι Boötis	5.3-6	261 "	White and ash color.
μ² (*Mu²*) Boötis	6.5-8	280 "	Both white.
Σ 1757 Virginis	8-9	292 "	White and yellow.
36 Andromedæ	6-7	316 "	Orange and yellow.
δ (*Delta*) Cygni	2.9-8	336 "	White and blue.
Σ 1819 Virginis	7-8	380(?)"	Both white.
μ (*Mu*) Draconis	5 5	643 "	Both white.
12 Lynx (trinary) A B	5.3-6.5	676 "	White and reddish.
ζ (*Zeta*) Aquarii	3.5-4.4	800(?)"	White and green.
Castor	2.5-2.8	1,000 " (about)	Both white.

FINEST COLORED DOUBLE STARS.

STARS.	MAGNI-TUDES.	DISTANCES.	COLORS.
γ (*Gamma*) Andromedæ...	2.2–5.5–6.5	10″–0″.5	Orange, sea green and blue.
Cor Caroli II......................	3.2–5.7	20″	Gold-yellow and lilac.
β (*Beta*) Cygni.........	3.4–6.0	34″	Gold-yellow and sapphire.
ε (*Epsilon*) Boötis...............	2.4–6.5	2″.9	Gold-yellow and blue.
95 Herculis.......................	5.5–5.8	6″	Gold-yellow and azure.
α (*Alpha*) Herculis..............	4 var.–5.5	4″.7	Orange and emerald.
γ (*Gamma*) Delphini............	3.4 6.0	11″	Orange and green.
32 Eridani.......................	4.7–7	6″.7	Topaz and marine blue.
ε (*Epsilon*) Hydræ	3.5–7.5	3″.5	Yellow and blue.
γ (*Gamma*) Ceti	3.2–7	3″	Pale-yellow and blue.
ζ (*Zeta*) Lyræ.........	4.5–5.5	44″	Yellow and green.
ι (*Iota*) Cancri	4.5–7	30″	Pale-orange and blue.
6 Triangul.........	5.5–6.5	3″.7	Gold-yellow and bluish-green.
Antares	1.7–7	3″.3	Orange and green.
ο (*Omicron*) Cygni..............	4.3–7.5–5.5	1′ 47″–5′ 38″	Yellow, blue and blue.
24 Comæ Berenices..............	5.6–7	21″	Orange and lilac.
ο (*Omicron*) Cephei.............	5.4 8	2″.5	Gold-yellow and azure.
94 Aquarii........	5.5–7.5	14″	Rose and light blue.
39 Ophiuchi.....................	5.7–7.5	12″	Yellow and blue.
17 Virginis.....	6.5–9	20″	Rose and red.
84 Virginis.......................	5.8–8.5	8″.5	Yellow and blue.
41 Aquarii.....	5.8–8.5	4″.8	Topaz and blue.
39 A Eridani....................	5.2–9	6″.4	Yellow and blue.
2 Canum Venatici................	6.0–9	11″	Gold-yellow and azure.
52 Cygni.........................	4.6–9	7″	Orange and blue.
55 Piscium.......................	6.0–9	6″	Orange and blue.
54 Hydræ.........................	5.2–8	9″	Yellow and violet.
66 Ceti...........................	6.5–8	15″	Yellow and blue.
ψ (*Psi*) Draconis................	4.8–6	31″	Yellow and blue.
η (*Eta*) Cassiopeiæ...	4.0–7	6″.7	Gold-yellow and purple.
σ (*Sigma*) Capricorni............	5.7–10	54″	Orange and lilac.
ν (*Nu*) Ursæ Majoris.............	3.3–10	7″	Yellow and blue.
Rigel	1.0–9	9″.5	White and blue.
δ (*Delta*) Herculis	3.6–8	18″	White and violet.
ο (*Omicron*) Capricorni..........	6.3–7	22″	Both bluish.
17 Virginis.......	6.5–9	20″	Both rose.

FINEST WHITE DOUBLE STARS.

STARS.	MAGNI-TUDES.	DIS-TANCES.	STARS.	MAGNI-TUDES.	DIS-TANCES.
Mizar.....................	2.4–4.0	14″	γ (*Gamma*) Leonis........	2.5–4.0	3″
Castor	2.5–3.0	5″.6	β (*Beta*) Scorpionis.......	2.5–5.5	13″
γ (*Gamma*) Virginis......	3.0–3.2	5″	θ (*Theta*) Serpentis.......	4.4–5.0	21″
γ (*Gamma*) Arietis	4.2–4.5	8″.9	44 ι Boötis..............	5.0–6.0	4″.8
ζ (*Zeta*) Aquarii	3.5–4.4	3″.5	π (*Pi*) Boötis.............	4.3–6.0	6″

STARS FOR WHICH A PARALLAX HAS BEEN FOUND.

The following stars, not appearing in the description, have been observed for parallax by Mr. J. C. Kapteyn, at Leyden, in 1891. (Revue d'Ast., August, 1891.)

DESCRIPTION.	MAGNI-TUDE.	* PARALLAX.	PROPER MOTION.
18115 Lalande	7.4	$0''.074 \pm 0''.027$	$1''.69$
θ (*Theta*) Ursæ Majoris	3.0	$0''.052 \pm 0''.026$	$1''.11$
19022 Lalande	8.1	$0''.064 \pm 0''.022$	$0''.79$
20 Leonis Minoris	5.8	$0''.062 \pm 0''.029$	$0''.69$
1618 Groombridge	7.0	$0''.176 \pm 0''.024$	$1''.43$
1648 Groombridge	6.3	$0''.101 \pm 0''.026$	$0''.89$
P. X. 96	7.4	$0''.038 \pm 0''.027$	$0''.27$
1812 Groombridge	6.7	$0''.030 \pm 0''.027$	$0''.64$
1822 Groombridge	8.0	$0''.016 \pm 0''.032$	$0''.67$
1855 Groombridge	7.3	$0''.056 \pm 0''.034$	$0''.33$

* See the note page x.

STARS OF GREATEST PROPER MOTION.

1830 Groombridge in Ursa Major	7″.03 per year.
9352 Lac. in Piscis Notius	6″.96 "
61 Cygni	5″.08 "
21185 Lal. in Ursa Major	4″.69 "
ε (*Epsilon*) in Indus	4″.67 "
μ (*Mu*) Cassiopeiæ	4″.43 "
21258 Lal. in Ursa Major	4″.37 "
o² (*Omicron²*) in Eridanus	4″.10 "
α (*Alpha*) Centauri	3″.64 "

1830 Groombridge travels 6,425 million miles per year or 1,760,000 miles per day.

o² Eridani	"	1,700	"	"	465,700 "
61 Cygni	"	925	"	"	253,400 "
α Centauri	"	370	"	"	101,400 "

These figures are only minima!

SHOOTING STARS—STAR-SHOWERS.

The shooting stars are small cosmical bodies appearing suddenly in some portions of the sky, leaving sometimes, for a few seconds, a luminous train. They become visible by coming in contact with the upper part of our atmosphere, and entering it with a velocity of from 10 to 50 miles per second. The systematical observations of the shooting stars are very recent. Palmer, after the extraordinary star-shower of November 12th, 1833, remembered a similar apparition observed on November 12th, 1799, by Humboldt. November 13th, 1866, a very remarkable shower also appeared. The period is 33 years.

Mr. Schiaparelli, taking into consideration the great velocity of these meteors, thought that they were circulating, like the comets, in regular orbits around the sun. He calculated the paths of the shooting stars of the 10th of August and of the 13th of November, and found that they were corresponding exactly with the orbits of two well-known comets — the one of the 10th to 11th of August with the comet of 1866, and the other with the great comet of 1862.

In 1872 the earth crossed the orbit of Biela's comet the 27th of November; a splendid star-shower happened that day. The 27th of November, 1885, another one appeared.

Mr. Coulvier-Gravier noticed that the shooting stars were more numerous in the morning than in the evening, and more in autumn than in spring; also more in the east than in the west. The reason for this was given by A. Herschel and H. Newton as the result of the double motion of the earth: its rotation and its annual motion around the sun.

These star-showers seem to come from some particular portions of the heavens called the *radiant points*. Mr. Heis is the first one who published a catalogue of these points, and since, many others, among them A. Herschel, Tupman and Denning, in England; Schmidt, in Greece; Schiaparelli, in Italy, etc., added their own observations.

Though every year does not bring the same phenomena, still we will give the dates and the radiant points of the principal showers, as they are capable, at any time, of offering a grand spectacle to observers:

IN FEBRUARY.—16th, near α (*alpha*) Aurigæ.

IN MARCH.—7th, near β (*beta*) Scorpionis and γ (*gamma*) Herculis.

IN APRIL.—9th, near π (*pi*) Herculis; 16th to 30th, near η (*eta*) Bootis; 19th to 30th, near 104 Herculis; 29th to May 2d, near α (*alpha*) Aquarii.

IN MAY.—22d, near α (*alpha*) Coronæ Borealis.

IN JULY.—23d to 25th, near β (*beta*) Persei; 25th to 28th, near ι (*iota*) Pegasi; 26th to 29th, near δ (*delta*) in Piscis Notius; 27th, near δ (*delta*) Andromedæ; 27th to 29th, near δ (*delta*) Aquarii; 31st, near α (*alpha*) Cygni; 27th to August 4th, near β (*beta*) in Triangulum.

IN AUGUST.—7th to 11th, near χ (*chi*) Cygni; 7th to 12th, near δ (*delta*) Dragonis; 8th to 9th, near α (*alpha*) Cassiopeiæ; 9th to 11th, near η (*eta*) Persei. (This star-shower is sometimes beautiful, and is known as the Perseïds.) 9th to 14th, near β (*beta*) Ceti; 12th to 13th, near 3084 Bradley; 12th to 16th, near μ (*mu*) Persei. (The shooting stars of this part of August are named by the French "Larmes de St. Laurent"— St. Lawrence's Tears.) 20th and 25th, near μ (*mu*) Persei; 21st to 23d, near o (*omicron*) Draconis; 23d to September 1st, near α (*alpha*) Lyræ; 25th to 30th, near η (*eta*) Draconis.

IN SEPTEMBER.—3d, near 14 Andromedæ; 3d to 14th, near β (*beta*) and γ (*gamma*) in the constellation Pisces; 6th to 8th, near ε (*epsilon*) Persei; 8th to 10th, near ζ (*zeta*) Tauri; 13th, near 236 Piazzi IV*h*; 15th to 20th, near β (*beta*) Andromedæ; 15th and 22d, near γ (*gamma*) Pegasi; 20th to 21st, near 42 Camelopardali; 21st to 22d, near α (*alpha*) Aurigæ; 21st and 24th, near β (*beta*) in Triangulum; 21st, near α (*alpha*) Arietis; 20th to October 9th, near γ (*gamma*) Arietis.

IN OCTOBER.—7th, near α (*alpha*) Arietis; 8th, near η (*eta*) Persei; 19th to 25th, between α (*alpha*) and β (*beta*) Tauri, near γ (*gamma*) Geminorum, and near Pollux.

IN NOVEMBER.—(The principal star-shower appears on the night of the 13th to the 14th, near κ (*kappa*), in the constellation Leo; it is for that reason called Leonids. The maximum of intensity happens every 33 years; the next will very likely occur again in 1899.) 1st to 8th, near A, Tauri; 16th and 25th to 28th, near μ (*mu*) Ursæ Majoris; 20th and 27th, near ω (*omega*) Tauri; 27th, near γ (*gamma*) Andromedæ (this shower is connected with the Biela's comet); 28th, near α (*alpha*) Cephei.

IN DECEMBER.—1st, near η (*eta*) Persei; 1st to 10th, between α (*alpha*) and β (*beta*) Geminorum; 6th, near ζ (*zeta*) Tauri; 6th to 13th, near 254 Piazzi IX*h*, α (*alpha*) Geminorum, and ι (*iota*) Ursæ Majoris.

According to Prof. Simon Newcomb, the average number of shooting stars falling on the earth is no less than 146,000,000,000 (146 billions) per year!

It is becoming generally admitted that aërolites, *bolides* and shooting stars are of the same origin as comets.

COMETS.

Of all heavenly bodies none attract the attention of the public so much as the large comets; "their rarity, their sudden apparition and their mysterious aspect astonish the most indifferent." It is but recently that their orbits have been calculated, and the return of Halley's comet was a complete success in the history of astronomy, in proving that like the planets and all other celestial bodies they obey the great law of gravitation.

The comets cannot be recognized by their appearance; they change sometimes very quickly, and it is only by comparing the elements of their orbits that they can be identified. When they follow an ellipse they are periodical; when their path is a parabola they are only visitors of our system for the time being and disappear forever.

"In common language, a comet consists of *head* and *tail*," said the Rev. T.W. Webb, but for astronomers it is not exactly so. The brighter part is called the *nucleus*, the mist around the nucleus is called the *coma*, and the nucleus and the coma form what is generally called the *head*. The *tail*, which is merely a prolongation of the head, *is always in the opposite direction of the sun* at the time of observation; it is sometimes straight, sometimes curved, sometimes divided in two, three or more branches, but most of the comets that become accessible to the telescopes are only faint, filmy masses, without heads and tails.

Astronomers are yet at a loss as to the nature and composition of comets. They come from all parts of the heavens, move in every possible direction, and but very few are visible to the naked eye. No year passes by without the discovery of several, and Kepler said that the ocean was not fuller of fishes than the æther of comets. Babinet called them "*des riens visibles*" (visible nothings), and John Herschel, in "Outlines of Astronomy," said that the tail of a large comet may weigh only a *few pounds* and perhaps only a *few ounces*. Stars of very small magnitude are often seen through the tail and even the coma of the comets without any sensible change in their brightness.

The principal periodical comets observed are:

1.	Encke........	period, 3 years, 105 days.		6.	Biela..........	period, 6 years, 220 days.			
2.	Tempel I........	"	5	"	73	"	7.	D'Arrest.......	" 6 " 235 "
3.	Brorsen..........	"	5	"	169	"	8.	Faye...........	" 7 " 207 "
4.	Winnecke.......	"	5	"	269	"	9.	Tuttle..........	" 13 " 296 "
5.	Tempel II........	"	6	"	....		10.	Halley..........	" 76 " i

In 1846 Biela's comet was divided in two parts and since then no trace of it has been found. The 27th of November, 1872, a beautiful star-shower was seen and identified with the comet Brorsen's comet should have returned in 1890, but was not seen. It is probably lost.

August 1st, 1889, Mr. Barnard discovered four satellite comets accompanying Brook's comet V. 1889.

According to M. Liais the earth and the moon passed through the tail of the great comet of 1861, on the 30th of June of that year.

The comets having the longest tails which have been calculated are the great comets of

1843,	which had a tail of.....	200,000,000 miles.		1858,	which had a tail of.....	55,000,000 miles.		
1680,	" " "	....150,000,000 "		1618,	" " "	.... 50,000,000 "		
1847,	" " "	....132,000,000 "		1861,	" " "	.... 42,000,000 "		
1811,	" " "	....111,000,000 "		1769,	" " "	.... 40,000,000 "		
1882,	" " "	.... 70,000,000 "		1860,	" " "	.... 22,000,000 "		
*1887,	" " "	.... 70,000,000 "		1744,	" " "	.... 17,500,000 "		

*This comet was only visible a few days in the Southern Hemisphere from the 18th to the 29th of January, length of the tail 30°.

THE PLANETS.

The planets, being celestial bodies revolving around the sun in regular orbits, can not be described among the constellations, as their position *on the sky* changes every day, but as their location is indicated in most every almanac for every month of the year, with a little attention they can not be mistaken for stars.

The moon is certainly the easiest object to observe with a telescope; its remarkable surface, its numerous craters, the shadow of its mountains, are worth seeing at any time, but more especially near the first quarter.

The sun is a little more difficult to observe, but sometimes the spots on its surface are large enough to be seen with the naked eye.

Mercury is only visible as morning or evening star, and never appears farther than 28 degrees from the sun, consequently it is difficult to be seen, as it appears near the horizon no more than two hours before sunrise or two hours after sunset. With the aid of the telescope it offers same phases as the moon.

Venus is the brightest planet, brighter than Sirius, and, like Mercury, *never passes the meridian at midnight;* it is morning or evening star, and its greatest elongation from the sun never exceeds 48 degrees; the smallest telescope will show the phases of Venus. When seen for the first time by a student, it may be easily mistaken for the moon; it shines sometimes four hours before sunrise and sometimes four hours after sunset.

Mars appears like a red star of the first magnitude, and its color will help one to recognize it easily among the constellations.

Between Mars and Jupiter there are now (February, 1892) 323 minor planets, or asteroids, two or three only being hardly visible to the naked eye when their position is known. The first one, Ceres, was discovered by Piazzi, January 1st, 1801; the 323d by M. Max Wolf. (Revue d'Ast., February, 1892).

Jupiter appears as a beautiful star of the first magnitude; its *belts* and its four satellites are easily seen with common telescopes.

Saturn appears as a dull star of the first magnitude; its rings and its largest satellite are accessible to ordinary telescopes.

Uranus, discovered by W. Herschel in 1781, is visible to the naked eye as a star of the sixth magnitude only, and difficult to be found without knowing its exact position.

Neptune was discovered by Le Verrier in 1846, and by Adams in the same year, but as the latter made his calculations public after Le Verrier's, the French astronomer has the glory of its discovery; it is never visible to the naked eye, and appears as a star of the eighth magnitude.

TABLE OF THE PRINCIPAL, ELEMENTS OF THE SOLAR SYSTEM.

Names and Signs.	Distance from the Sun. Earth=1.	Distance from the Sun in miles.	Diameter, Earth=1.	Diameter in miles.	Area. Earth=1.	Volume. Earth=1.	Time of Rotation.	Time of Revolution.	Weight. Earth=1.	Density. Earth=1.	Density. Water=1.	Light and Heat received from Sun. Earth=1.	Apparent Diameter of Sun from the Planets.	Apparent Diameter of the Earth from the Planets.
☉ Sun			108.556	845,000	11,450.000	1,279,267,130	612 0 0	Unknown.	324,479,000	0.25	1.38			
☿ Mercury	0.387	35,750,000	0.376	3,060	0.148	0.050	*24 5 28	87 23 14	0.075	1.38	7.50	6.674	1° 21'	20''
♀ Venus	0.723	66,720,000	0.964	7,500	0.925	0.868	*23 21 24	224 16 41	0.787	0.98	5.04	1.911	43'	65''
⊕ Earth	1.000	92,000,000	1.000	7,950	1.000	1.000	23 56 4	365 5 48	1.000	1.00	5.48	1.000	30'	
♂ Mars	1.524	140,750,000	0.540	4,210	0.290	0.157	24 37 23	1 321 22 18	0.109	0.71	3.94	0.431	21'	1° 54''
Asteroids (average)	2.500	250,000,000	0.050	400	0.003	0.001		4 255				0.140		58''
♃ Jupiter	5.203	481,250,000	11.160	83,000	125.000	1,234.000	9 55 45	11 315 12	300.028	0.24	1.31	0.037	6'	Little Star.
♄ Saturn	9.539	887,500,000	9.527	76,320	90.000	864.690	10 16	29 167 4	91.931	0.13	0.76	0.011	3' 22''	Very little Star
♅ Uranus	19.183	1,775,000,000	4.221	34,500	18.000	75.250		84 89 9	15.771	0.12	0.98	0.003	1' 40''	Invisible.
♆ Neptune	30.037	2,750,000,000	4.407	37,500	22.000	85.600		164 226	18.542	0.22	1.21	0.001	1' 4''	Invisible.

* From recent careful observations of Schiaparelli, Mercury and Venus accomplish their rotations *in the same time as they do their revolutions;* they turn around the Sun as the Moon does around the Earth. This remarkable discovery is yet open to discussion.

Some astronomers believe in the existence of another planet circulating between the Sun and Mercury, but all efforts to see it during the total eclipses of the Sun have failed.

The *transits* of Mercury happen frequently. The last one occurred May 10th, 1891; the next will be November 10th, 1894, November 12th, 1907, November 6th, 1914, May 7th, 1924, November 8th, 1927, May 10th, 1937, November 12th, 1940, etc.

Venus is the most difficult planet to be well defined with telescopes. The best time of observation is through the day. The transits of Venus occur *every one hundred and thirteen and one-half years plus or minus eight years.* The last ones occurred June 5th, 1761, June 3d, 1769, December 8th, 1874, December 6th, 1882; the next will be June 7th, 2004 and June 6th, 2012.

The Moon revolves around the Earth at a distance varying between 252,972 and 221,614 miles. The eclipses of the Moon and the Sun are always interesting, and should not be missed by amateurs.

Mars is the planet having the most points of resemblance to the Earth, but it requires a good telescope to see it. Its two satellites were discovered by A. Hall, at Washington, in 1877. They are supposed to be only six or seven miles in diameter.

†Jupiter is a beautiful object for the telescope. An ordinary instrument will show it as an imperfect circle, a little flattened at the poles. The eclipses and transits of the four satellites are very frequent and always interesting. Their diameters are about 2,500, 2,100, 3,550 and 2,960 miles. W. Herschel and Schræter think that they always show the same side to Jupiter, as the Moon does to the Earth.

Saturn, the wonder of the Solar System, is also accessible to common telescopes. The revolution being nearly twenty-nine and one half years, its system of rings increase and decrease in breadth in a little less than fifteen years. The wide opening happened in 1885-86, and this year (1892) we will see one its edge, they will keep increasing until 1899. Besides these rings Saturn has eight satellites, the largest, Titan, appearing as a star of the 8.5 magnitude, is larger than Mars; Japet, or Iapetus, is about the size of Mercury; Rhea, about the size of the Moon, and the other five have diameters from 450 to 1,250 miles.

Uranus has four satellites and Neptune one. Some perturbations of this last planet lead to the conclusion that there is another planet beyond.

† Prof. Barnard, of the Lick Observatory, discovered on September 9th, 1892, a new satellite of Jupiter, shining as a star of the 13th magnitude, revolving around its primary in about 12 hours and 60 minutes at a distance of 112,460 miles from its center.

In conclusion, we desire to say that our Handbook is not intended for theoretical astronomy, but merely for the general public who wish to obtain a fair idea of the beauty of the heavens without great expense of time and money, and in its compilation we have drawn from the works of S. W. Burnham, H. Faye, C. Flammarion, R. A. Proctor, T. W. Webb and other noted astronomers, also from the Revue d'Astronomie Populaire, published monthly since 1882.

We desire especially to extend our sincere thanks to Mr. Burnham for valuable information and also for corrections on the advance proofs, which were submitted to him. If by this little work we have encouraged our readers to pursue further this beautiful study our object is attained.

INDEX OF ILLUSTRATIONS.

Illustrations of the Principal Double and Triple Stars, Binaries, Trinaries, etc.

Illustrations of the Principal Clusters and Nebulæ.

Diagrams, Orbits of Stars, etc.

GENERAL INDEX.

www.ingramcontent.com/pod-product-compliance
Lightning Source LLC
Chambersburg PA
CBHW030624270326
41927CB00007B/1292